Bringing Technology to Market

1807 – 2007 Knowledge for Generations

Each generation has its unique needs and aspirations. When Charles Wiley first opened his small printing shop in lower Manhattan in 1807, it was a generation of boundless potential searching for an identity. And we were there, helping to define a new American literary tradition. Over half a century later, in the midst of the Second Industrial Revolution, it was a generation focused on building the future. Once again, we were there, supplying the critical scientific, technical, and engineering knowledge that helped frame the world. Throughout the 20th Century, and into the new millennium, nations began to reach out beyond their own borders and a new international community was born. Wiley was there, expanding its operations around the world to enable a global exchange of ideas, opinions, and know-how.

For 200 years, Wiley has been an integral part of each generation's journey, enabling the flow of information and understanding necessary to meet their needs and fulfill their aspirations. Today, bold new technologies are changing the way we live and learn. Wiley will be there, providing you the must-have knowledge you need to imagine new worlds, new possibilities, and new opportunities.

Generations come and go, but you can always count on Wiley to provide you the knowledge you need, when and where you need it!

William J. Pesce

William J. Pesce
President and Chief Executive Officer

Peter Booth Wiley

Peter Booth Wiley
Chairman of the Board

Bringing Technology to Market

Trends, Cases, Solutions

Edited by
Olaf Plötner and Robert E. Spekman

BICENTENNIAL 1807 WILEY 2007 BICENTENNIAL

WILEY-VCH Verlag GmbH & Co. KGaA

Library of Congress Card No.:
applied for

British Library Cataloging-in-Publication Data:
A catalogue record for this book is available from the British Library.

Bibliographic information published by the Deutsche Nationalbibliothek
The Deutsche Nationalbibliothek lists this publication in the Deutsche Nationalbibliografie; detailed bibliographic data are available in the Internet at http://dnb.d-nb.de.

Printed in the Federal Republic of Germany

Printed on acid-free paper

Typesetting Mitterweger & Partner, Plankstadt
Printing and binding Ebner & Spiegel GmbH
Cover Design init GmbH, Bielefeld

ISBN 978-3-527-50270-7

Table of Contents

Bringing Technology to Market. Olaf Plötner and Robert E. Spekman

ISBN 978-3-527-50270-7

Introduction

Olaf Plötner

This book was written for the B2B community on both sides of the Atlantic; we have met many of its members during various programs treating different aspects of bringing technology to market. The chapters that follow reflect our own explorations in this area as well as those of students and B2B practitioners who have shared their experiences with us. As a result of such discussions, we realized we should devote as much space in these pages to practical as to theoretical topics; the reader will thus find chapters that consider a range of aspects of current industrial marketing theory and others offering support in the form of real-life cases. For clarity the theoretical chapters are grouped around *external* aspects of strategy on the one hand and *internal* questions of organizational systems on the other hand.

This books begins where strategic concerns begin today, namely with global markets and the drivers of the competitive forces resulting from them. As Albrecht Söllner and Wulff Plinke observe in their contribution, both new sales possibilities and risks tied to intensified competition continue to emerge from these markets. The case of the ICT house T-Systems highlights the decision-making processes at play here, supplying an overview of goal setting and subsequent system alignment in the areas of region, customers, and organization.

In the larger economic context, the forces at work in global markets require even greater alignments. This is especially true when new strategic starting bases lead to a disruption of traditional value chains and the formation of previously unfamiliar ones. Michael Kleinaltenkamp traces the backdrop to this development, which we illustrate with the case of MAN Roland printing machines. "The situation in the printing industry remains tense worldwide.", states the company's annual report for 2005. "The sector has been characterized by continuing concentration, plus globally fierce competition between printing press manufacturers. Against this background MAN Roland succeeded in winning a highly attractive order by offering extensive services." Such service agreements had been unusual in the printing industry; they represented a new business model, with downstream activities previously covered by the client now turned into an element of the supplier's value chain, one eventually maturing into a unique selling proposition.

At present, many traditionally goods-oriented B2B enterprises view the forced expansion of the service business in a highly positive light: as an opportunity to gain global competitive profile and move toward achieving ambitious growth targets. To be sure, the new service offerings do not always produce the

Bringing Technology to Market. Olaf Plötner and Robert E. Spekman

ISBN 978-3-527-50270-7

desired economic success. In chapter 2 we thus examine the specific challenges that need to be met in marketing B2B-connected services. That discussion is supplemented by a comparative case study of Siemens and IBM – two IT-industry firms engaged over a number of years in the shift from a goods-oriented to a service-oriented business. In this regard, the differences in developmental results emerging from each company's specific structural conditions and strategic management decisions is striking.

This theme is examined in an operative framework in our case study of Leistritz AG, with special focus on the question of the ways in which suppliers can convince buyers of the benefits of innovative service-packages. This touches on the connection between the perspectives of pricing and costing – a topic explored more deeply in the contribution of Mario Rese.

B2B business has always been characterized by the high complexity of the products and services to be sold. Marketing services requires both competent managers and an excellent sales organization. In their chapter, Wesley J. Johnston and Brian N. Rutherford offer an overview of the rules and regulations at play in this business realm, while Jagdish N. Sheth and Arun Sharma address the question of the increasing globalization of the B2B business. An additional contribution by Olaf Plötner and Frank Jacob discusses communication policies in terms of both goals and resulting organizational forms.

With Siemens SE&A, an example is here offered of a sales organization that serves as a benchmark for many companies. In the authors' words, this firm has furnished "the groundwork for a consistent business planning process that ultimately results in business target agreements between the global and regional units and between sales and marketing."

In the traditional sales force, individuals usually brought product focus, transactional knowledge, and persuasive skills together for the sake of closing a sale. At present, this set of qualities is widely viewed as rather crude or inadequate, hence as counterproductive when it comes to establishing long term client-relationships. Today's sales executives simply need a much wider range of skills than in the past; consequently, in the realm of key account management, they have now taken on the role of general managers – individuals possessing the requisite cultural experience for dealing with global partners. A company must thus have a solid sense of which personnel to choose for positions demanding technical, social, practical, and political expertise. Nevertheless, organizing a sales force still first and foremost means taking care of the basics, which is to say following the rules of efficiency and effectiveness (the traditional levers of product and territory management) as well as arranging task-related, performance-based compensation. In their chapter, Robert E. Spekman and Benjamin Moor thus consider the special dynamic of the sales force bringing new technology to market, together with the impact effective sales leadership has in responding to challenges that are complex and global at the same time. As the authors describe these challenges, "the boundary-role individuals involved must manage across units and companies, forming teams whose members typically see themselves as individual contributors, and developing behaviors that support working with customers as trusted advisors and honest brokers."

Requirements such as these call for transformational leadership models tied to compensation plans not only matching the long selling-processes typical for B2B business but also complying with a system demanding knowledge of markets, products, and customers; relationship-building skills; global orientation; and leading qualities suitable for culturally diverse teams. The system chosen to control such a sales performance may be based on outcome or behavior or consist of a mix of the two. Due to the collaborative aspects of B2B deals, the latter is especially the case for the industrial goods and services industries. If we follow Martin Kupp in his contribution, the influence of "contextual factors" such as "environmental turbulence and selling complexity ... is even more important than salesperson characteristics" in choosing the proper control system. It is probably self-evident that such personal and managerial complexity requires a suitably adjusted organizational support system: one manifest not only within key account teams and marketing centers, but as an activity-based, customer-centered structure in which outside demands influence the internal organization rather than the other way around.

1
Market-Oriented Management in Global Markets

Albrecht Söllner
Wulff Plinke

The Effect of Globalization on the Market Process

Globalization can be understood as a process involving the expansion of the international division of labor. It is above all concerned with two domains: added value and sales. Through globalization, the possibilities for organizing added value are expanded, as are sales possibilities and the number of potential markets.

Globalization is not a process open to discussion. Rather, it is a fact that companies have to deal with if they are to remain competitive. Political decision making in most countries and technological progress are the driving forces behind globalization. One of the central insights gained from the great depressions of the 1920s and 1930s is that sealing off national economic areas with a policy of protectionism is no suitable response to economic problems. To the contrary: at the time, that generated stagnation and inefficiency. For this reason, most industrial countries and organizations such as the World Bank and the International Monetary Fund have committed themselves to a policy of maintaining and encouraging open domestic economies. In this framework, they have worked to do away with protectionism and reduce trade barriers. The lowering of import duties over recent years can be seen as a reflection of these policies.

But the consistent realization of an ideology of free trade is only possible because of technological advances. It is the case that intensified international division of labor does lead to a considerable rise in productivity, and with that to a lowering of production costs. Nevertheless, advantages in the latter area can be quickly balanced off – indeed even overcompensated for – by the heightened need for coordination at work in cross-border cooperation. The costs of such coordination (for example, the search for partners and their evaluation; negotiation and communication; adaptation and oversight of performance; realization of demands) are designated as transaction costs. They are just as real as production costs, even though their origins seem less tangible.

Over recent years, precisely transaction costs have fallen sharply as a result of technological progress. Performance advance in the information and communication technologies together with sinking costs for transport have allowed an international labor division of previously unimaginable scope. The cost of sea shipping alone has plummeted to a tenth of what it was before (see box below). The demand for transport capacity is so large that in specific areas it has caused bottlenecks and even increased prices.

Bringing Technology to Market. Olaf Plötner and Robert E. Spekman

ISBN 978-3-527-50270-7

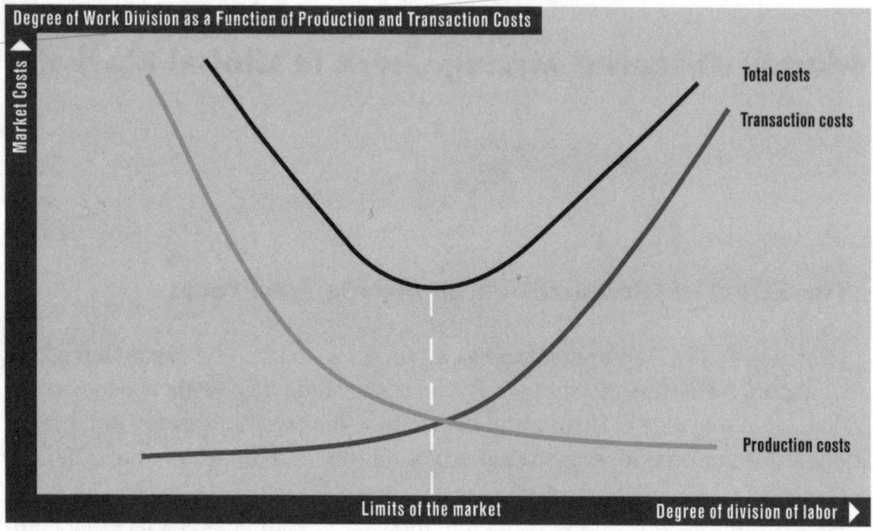

Fig. 1: Degree of Work Division as a Function of Production and Transaction Cost

Sinking fuel prices: Container ships as an example

Hardly any other branch of industry has profited from globalization as much as shipping. The shipping companies have been reporting record profits, the harbors record handling; financial sources for shipping such as banks are chartering out container giants for record sums; and the order books of the Asian dockyards are full. The worldwide fleet has nearly doubled since 1996, and its container capacity has expanded enormously. Now many of the giant ships cannot even pass through the Panama Canal.

Meanwhile nearly everything is transported with containers: Dutch tulip bulbs are shipped to New Zealand, crabs from the North Sea are shipped to Asia, and in Russia wood logs are automatically cut to container dimensions. The enormous growth in transport space has been made necessary by the increasingly networked economy and worldwide division of labor. And the prognosis is for a further increase in world trade – although the world's economy is stagnating.

The container ships play a decisive role in this boom in world trade. This is because of the lowered costs of transport, made possible through the growing capacity of the ships. These costs no longer carry much weight in production considerations. When for instance 15,000 bottles of Australian beer are shipped to Germany in a container, the cost amounts to less than five cents per bottle. Twenty years ago, the average quota of sea transport costs was around ten percent; today it is only around one percent. This is what makes the division of labor possible, since the cost advantage of producing goods in China, for example, only holds when transport is cheap. Container usage has been increased by China becoming a global factory and a serious exporter in its own right (in the past decade exports have quintupled). The exchange of goods is now taking place in both directions, no longer only from the Western industrial countries to Asia. This development has itself encouraged a demand for larger ships, the transport costs thus remaining low or sinking further, which in turn has a catalyzing effect on world trade.[1]

The increasing labor division and entwining of performance on an international level becomes apparent in a comparison of the growth of worldwide production with the growth of worldwide trade. The growth rates for trade exceed those for production by a wide margin (see fig. 2).

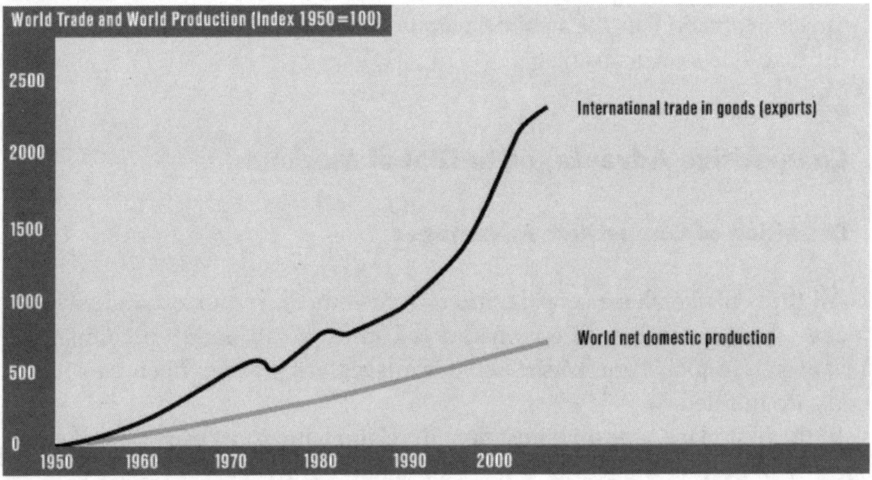

Fig. 2: World Trade and World Production (index 1950 = 100)[2]
Source: WTO, International Trade Statistics 2001

For individual companies, the expanded markets resulting from globalization have generated both risks and opportunities. New markets offer additional sales possibilities. At the same time the intensity of competition increases. This situation creates considerable challenges:

- The firms have to recognize the opportunities and risks of the expanded *market outlets*. New customers, but also new competitors in new or traditional markets, are dramatically altering the competitive situation. The company's search for its own position in an internationally unfolding marketing and competitive process stands here in the foreground. In this context of global competition, correctly evaluating customer problems has taken on an importance as great as that which it has had in domestic markets.

- Globally competing firms are being forced to fundamentally reshape their *value chain*. The challenge lies in the coordination of the production process across national borders. The following questions are pertinent here: What should the company make itself and what should it acquire elsewhere (make-or-buy)? From whom and how should this latter category of goods and services be acquired? How should both the high rate of R&D investment tied to technological advances and the high fixed-cost blocks be dealt with? How can scale economies be achieved? And how should the firm develop its own performance potential and resources? As core areas of the firm, management and the sales force – considered later in this book – are closely tied to this last question.
- But challenges emerge for a company not only in relation to target markets and arranging its own resources. To a large degree, company success depends on the rules that the marketing process follows. Recognizing both these rules (laws, regulations, and so forth) and developmental trends allows quick adaptation to the new developments, together with influencing the rules to one's own advantage.

Competitive Advantages in Global Markets

Definition of Competitive Advantages

All three of the above considerations – customers, resources, and environment – have an influence on whether a company can build up competitive advantages in global markets. Such advantages are present when two conditions are fulfilled:

In the first place, a positive cost-benefit relationship must exist from the perspective of each of the participating transaction partners.

In the second place, each particular cost-benefit bundle (each set) has to be perceived as superior to alternative choices.

This describes a first essential element of a competitive advantage. A second element involves the costs of a supplier A as compared to a competitor AC.[3] A relative cost advantage can be described as the difference between the cost of sales of supplier A and those of the relevant supplier AC.

A favorable cost position points to the superiority of supplier A. Although suppliers' costs do not necessarily have anything to do with the prices that customers pay – in many cases customers are unaware of these costs – an advantageous relative cost position offers substantial benefits. Through the higher productivity, the suppliers can also hand the cost advantage down to their customers in the form of lower prices, thus raising the customers' net benefit. An advantage of supplier A over competitor AC can be defined as follows:

Net benefit difference$^{A/AC}$ of the purchaser in favor of competitor (supplier)^{A}
+ cost difference $^{A/AC}$ of supplier^{A}
= competitive advantage A/AC of supplier A

In this way the competitive advantage is a two-dimensional quantity whose one dimension is defined by effectivity (net benefit difference) and whose other dimension is defined by efficiency (cost advantage).

Effectivity here represents an external measure of performance indicating the extent to which a firm does justice to the expectations and demands of its customers. The expectations do not regard a product but rather both a problem experienced by the customer and its solution, which includes solving problems of coordination. For its part, *efficiency* is an internal measure of performance allowing a depiction of the relation between output and input.[4] Effectivity and efficiency should here be understood in a broad sense. The main concern is virtually never only a product and its price but also the costs or benefits of coordination tied to a transaction.

Market Orientation as the Starting Point for Creating Competitive Advantages

Consistent market orientation represents a necessary response by firms to the increase of pressure in almost all markets in the course of globalization. At the same time, such an orientation is a precondition for creating competitive advantages. *Market orientation* simply refers to a particularly intensive focus on the market. It allows companies to produce services that solve the problems of target customers better than competitors. With their definition Narver and Slater make clear that market orientation includes an analysis of the competitors and the question of added value within the company. More precisely, they speak of "three performance components: customer orientation, competitor orientation, and interfunctional coordination; and two decision criteria: long-term focus and profitability."[5]

Three tasks for internationally operating companies can be derived from a market-oriented perspective:

(1) International firms have to examine whether their internal structures and processes are suited to offer their target customers a net benefit, and whether they can build up a defendable competitive advantage.
(2) International companies have to bring the type of coordination of the transaction with the customer into their analysis.
(3) The analysis needs to serve as a basis for ascertaining both the status quo of the company's own competitive position and the management implications for entering and handling the market.

Regarding (1): Companies distinguish themselves through company-specific structures and processes (resources). These resources determine how or whether a firm can satisfy customers' needs. A firm's share of advantageous resources is defendable – or, to use Ghemawat's term, "sticky" – when three conditions are fulfilled:[6]

The resources have to be *enduring*, hence capable of having a long-term impact. They need to be *special* in one way or another – in other words, distinguishable from factors that are usually in the picture – otherwise they would not have the potential to produce a competitive advantage. Finally, the resources must have limited *tradability*. If the resources could be traded without any problems, both sale and purchase would be easy, differences in competitive capacity thus quickly vanishing. Alongside the firm's resources, the prevailing conditions in which it operates can contribute to its success. This topic will be treated more closely below.

In an international framework, once resource advantages are present the question emerges of their transferability to new target markets. Although in recent years trade barriers and transaction costs have receded, cultural differences, for example, can represent market-entry barriers in various countries. In the international framework, consideration of the two dimensions of resource-advantage and the possibility of transferring that advantage allows a firm's first competitive positioning (see fig. 3).

Here, in each row of the matrix two results are distinguished according to whether a firm's assumed resource advantage can or cannot be transferred to

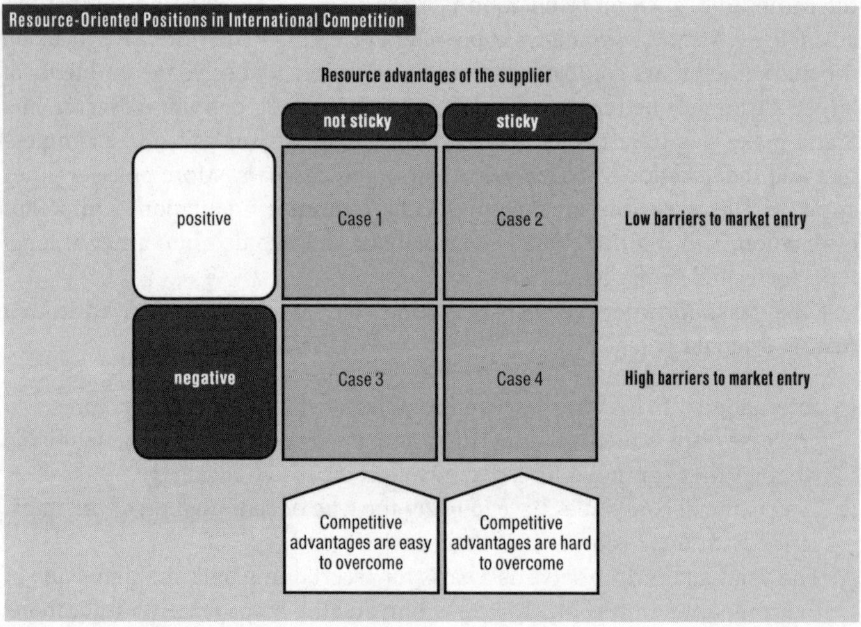

Fig. 3: Resource-Oriented Positions in International Competition

an external target market. In cases 1 and 2 that possibility is present; the barriers to market entry are thus relatively low. In cases 3 and 4 there are, to the contrary, no transferable resource advantages. In considering a potential target market, the company sees itself confronted with high barriers to market entry. The columns also contain a distinction between resource advantages that can be defended–hence are sticky–or not. Correspondingly, cases 1 and 3 refer to competitive advantages that are overcome by competing firms with relative ease. In contrast, for competitors the competitive advantages in cases 2 and 4 are only imitable with difficulty. The defensibility of competitive advantages is here not exclusively based on company resources. Special qualities of the country of origin – infrastructure, related and supportive industries, business clusters, etc.[7] – can lead to competitors with other origins only being able to imitate advantages with difficulty.

For a company that wishes to enter an external target market, case 2 is especially attractive. Market entry is here possible without problems and the company's own competitive advantage is defendable. In case 3, on the other hand, market entry is not advisable. Cases 1 and 3 do not offer a unified picture. In case 1, despite the ease of market entry, the danger exists that the prevailing competitive advantage will be quickly imitated. In case 4, despite such an advantage the company faces market-entry barriers.

Regarding (2): International firms have to bring the type of coordination of transaction with the customer into their analysis. Hence fig. 4 depicts a distinction between two types of transactions: From the customer's perspective, transactions can be either problematic or unproblematic. The source for problems here lies in the complexity or insecurity of a transaction; in insufficient knowl-

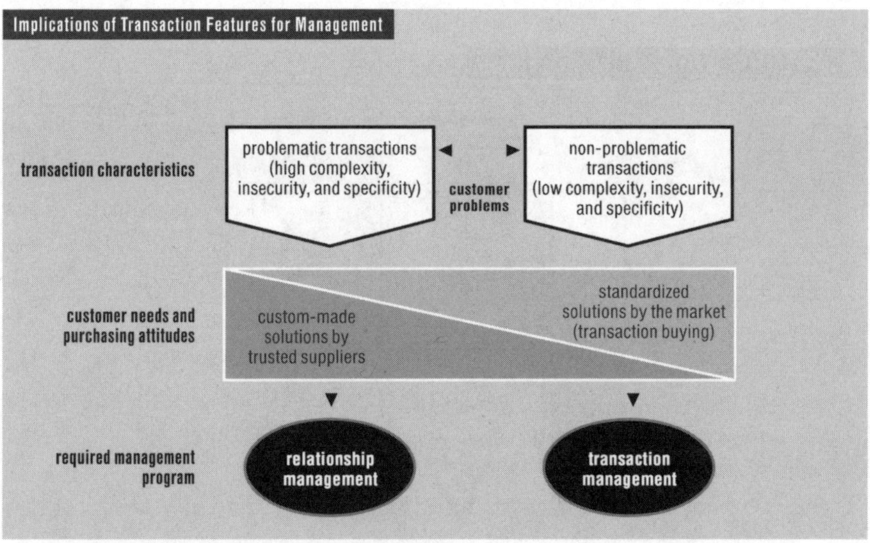

Fig. 4: Implications of Transaction Features for Management[9]
Source: Söllner: Die neue internationale labor division

edge of the product or desired solution by the customer; or again in forms of dependency towards the supplier resulting from a purchase.[8] In each of these cases, transaction costs in addition to the price of purchase – costs related to the searching, negotiating, and adjustment processes – are generated for the customer. The costs resulting from transactional problems have a profound influence on the customer's needs and purchasing attitudes. When it is possible to reduce the transaction costs by entering into a close business relationship, it can be highly efficient to limit one's own freedom of choice regarding a supplier and build close ties with one who is trustworthy. The customer will then prefer a custom-made solution from such a supplier to a standardized market solution.

We thus see that there are two types of customer and purchasing attitudes: On the one hand, there are customers who always newly arrive at their purchasing decisions, which is to say without any consideration of earlier decisions. These customers purchase products or services by means of the anonymous market (transaction buying). On the other hand, there are customers who repeatedly conduct transactions with the same business partner. In this case, marked by an inner linkage between the transactions, we can speak of relationship buying.[10]

For the supplier, highly different challenges result from transaction buying or relationship buying. Management of a business relationship poses other, higher demands than management of a discrete transaction. A business relationship cannot be built in a day; usually, it requires a considerable investment of time and resources. By developing a management program oriented around the customer's purchasing behavior, both bad investments and customer failures can be avoided (see fig. 5).

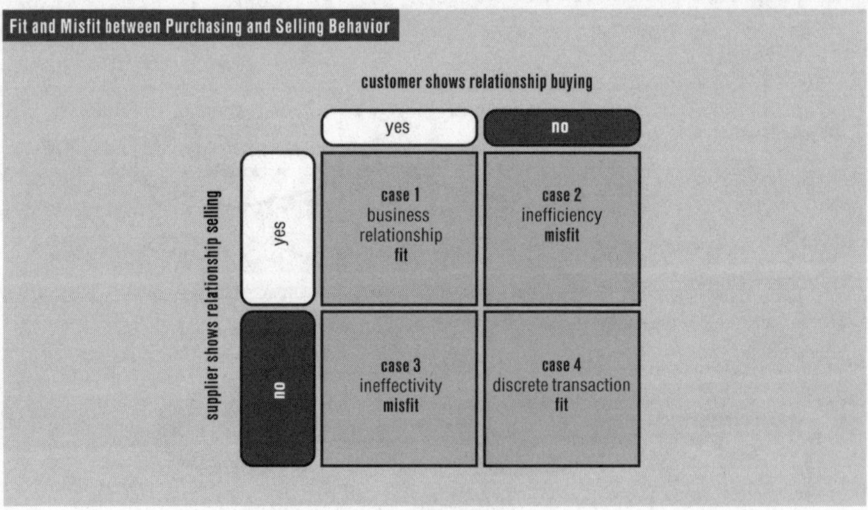

Fig. 5: Fit and Misfit between Purchasing and Selling Behavior[11]
Source: Plinke: Grundlagen des Businessbeziehungsmanagements

Altogether, four cases can be distinguished describing each particular fit or misfit between purchasing and selling behavior. In the first case – a fit – the customer desires a business relationship; that is, from the customer's perspective the transaction exhibits problematic characteristics. The supplier practices relationship selling. The behavioral program can be described as effective and efficient. It is effective because it solves the customer problem. And it is efficient because the supplier does not waste any funds. For potential competitors, the ties between supplier and customer represent a considerable (business-relationship) barrier. *Exchange barriers* result from the opportunity costs and relevant transaction costs of the exchange, together with the sunk, specific investments in the business relationship.[12]

In the second case the supplier again invests in constructing a business relationship. But this time, the customer fails to honor the effort. This misfit is inefficient for the supplier firm since it is wasting resources. In the third case the customer is in need of a trusting relationship with the supplier. This need is either not perceived or not fulfilled by the supplier, whose practical approach is thus ineffective. The fourth case again involves a fit: the two parties treat the transaction as a discrete transaction. A market transaction thus takes place that is not enriched with relational elements.

In discussing the problematics of a transaction, our focus has until now been on the customer's perspective. But in this respect problems possibly exist on the supplier's side as well, their causes the same as those found on the side of the customer. In international business, one special source of insecurity emerges from the danger that in case of a market entry, relevant knowledge will make its way to the competition. Such a danger is above all present when a firm chooses to enter the external market together with a partner, for example one from the target country itself.

Regarding (3): Management implications for market entry and development can be derived from an analysis of resource position and coordination needs.

Fig. 6 presents various conceivable positions in international competition. The perspective here is that of a firm that would like to assess its chances for entry into an external target market. The different cases addressed in the figure have far-reaching implications for the competitive process. Cases 1 to 4 refer to a fit between the attitudes of the current supplier and those of his customer. Cases 1 and 2 involve transaction buying and selling. In cases 3 and 4 supplier and customer initiate a business relationship. Hence in cases 1 and 2 an external competitor does not encounter existing business barriers. Especially the sticky resource advantage of the competitor in case 1 lays the groundwork for a lasting, defendable competitive advantage. In cases 3 and 4 the challenger is faced with relational barriers. In these cases, even with superior resources the supplier cannot count on quick success. Especially in those cases in which the challenger's resource advantage is not sticky, through the customer's exchange barriers the established supplier has a great deal of time to overcome that advantage.

Cases 5 to 8 involve misfits. In cases 5 and 6 the customer does not find the transaction problematic. Nevertheless, the established supplier has invested in developing a business relationship, thus threatening efficiency. In these cases, thanks to superior resources it should not be so difficult for a challenger to win customers in the new target market. In cases 7 und 8, the customer would gladly possess ties to selected suppliers in order to manage complex or risky transactions. But the established supplier does not know this. This constellation produces an opportunity for the challenger. At the same time, under such circumstances it may be much more difficult to achieve success. Building a business relationship demands investments and time. But above all, an external company has a starting disadvantage in comparison to an internal company; this is grounded in the special and sometimes cultural distance between supplier and consumer. Business relationships are meant to reduce insecurity in problematic transactions. Spatial proximity and mutual acquaintance is here an advantage that an external challenger may well not enjoy. This disadvantage can even outweigh the challenger's resource advantage.

From the supplier's view, the question also emerges of the extent to which acting alone is called for – or else handing certain activities to a partner. Cooperation is in any event only feasible when it does not lead to relevant knowledge being passed to a partner or a competitor. Cases 1, 3, 5, and 7, in which a defendable (sticky) advantage on the part of the challenger is present, are thus divided into sub-cases (a) and (b). Sub-case (a) suggests that competitive advantages are not endangered by cooperation, competition-relevant knowledge thus not being lost. Sub-case (b) indicates that cooperative knowledge cannot be protected, competitive disadvantages through knowledge outflow thus posing a danger in the long term.

But fig. 6 is not only suggestive in regards to a company's competitive situation. It also points to concrete recommendations for action regarding market entry and the development of international transactions (see fig. 7).

Positions in International Competition (Orientation toward Resources and Demand)

TB = Transaction buying
TS = Transaction selling
RB = Relationship buying
RS = Relationship selling

		Resource advantage of the challenger is		
		sticky		not sticky
		stickiness is not endangered by cooperation	stickiness is endangered by cooperation	
with entry into an external market, the challenger encounters a	fit between the purchasing attitudes of the customer and sales attitudes of the established supplier	1a TB/TS	1b TB/TS	2 TB/TS
		3a RB/RS	3b RB/RS	4 RB/RS
	misfit between the purchasing attitudes of the customer and sales attitudes of the established supplier	5a TB/RS	5b TB/RS	6 TB/RS
		7a RB/TS	7b RB/TS	8 RB/TS

Fig. 6: Positions in International Competition (Orientation toward Resources and Demand)

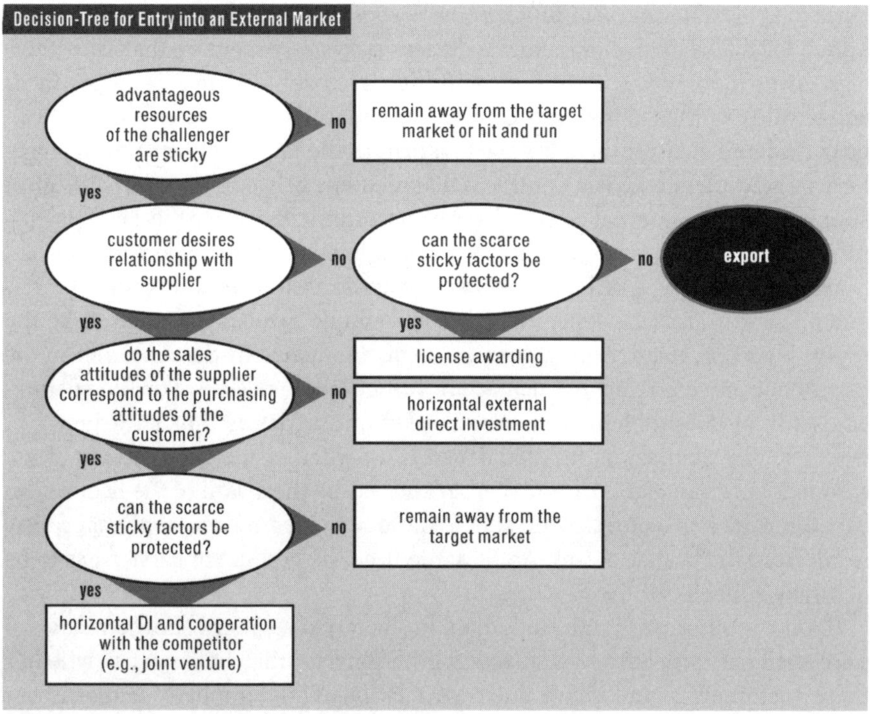

Fig. 7: Decision Tree for Entry into an External Market

From the challenger's perspective the first thing to ask is whether the advantageous resources that are available are also defendable; if they are not so, the expensive process of market entry could in any case not be considered on account of the danger of imitation. The message for the challenger then amounts to: stay away from the target market or embrace hit-and-run policies – which is to say, only enter the market when a faster profit is feasible and withdraw when the competition position worsens through imitation or another factor. To be sure, the possibility of quick market entry and exit is only present in those cases where the customer considers the transaction unproblematic.

If the challenger's resources are sticky, both the form of market entry and the future international labor division depends in a decisive way on the purchasing attitude of the customer. When a transactional attitude is apparent, the export business and license awarding are the two basic forms of market entry that come into question. *Export* here refers to the distribution of goods manufactured in a firm's home country in external target markets.

An export business, in the sense of the above-described transactional typology, is a type of business characterized by a lack of transactional problems from the customer's perspective, while from the supplier's perspective keeping control of the transaction seems called for. Since the customer does not need a close business relation, it is not necessary to be present on site and to try for close ties. Flexible market entry through export thus represents an interesting option for the challenger. It is markedly less risky and expensive than direct foreign investment while nevertheless fulfilling customers' needs. It produces added value in the challenger's country. As mentioned, an alternative that can be considered is awarding a license to a firm in the target country. Licensing is here to be understood as a contractual agreement between an internal license supplier and an external licensed party, according to which the latter is furnished with the disposition rights over intangible assets for a specified period.

Licensing as a type of business transaction derived from a theoretical referential frame should be separated from the simple licensing process. Like the export business, the licensing business is distinguished by the customer in the transaction *not observing any significant transactional problems.* For the customer's sense of the problem, a discrete market transaction is thus sufficient. But whether this approach is feasible depends on whether the defensibility of the competitive advantage is diminished in any way by the award of the license. No damaging loss of control over added value activities by the challenger may result from the license award. At the same time the product or service has to be naturally suitable for the licensing.

If, on the other hand, the customers in the target segment show a relational purchasing attitude, the next question emerging for the challenger is whether these customers already have business relations with suppliers. If that is not the case, suppliers have to direct their efforts at building up business relations. But this is hardly possible through market entry per export. What is here in question is above all the transactional type of foreign direct investment. Direct investments are cross-border investments aimed at an enduring influence on a company in another land.[13] Spatial proximity to the customer must be considered an important condition for relationship selling. Through a defendable resource advantage and the fact that the established suppliers have not yet developed business relations with the target customers, direct investment offers the challenger a possibility of successful market entry.

The case is completely different when target customers are already tied to a business relationship. Here the challenger cannot count on fast success. Even with market entry through direct investment, the customers are not ready to abandon the existing business relationship. To this extent, the question arises of whether a form of horizontal cooperation with the competitor is conceivable, for instance in the form of a joint venture. An international joint venture is an enterprise steered by two or more partners, with at least one of the partners settled abroad. Usually the participating parties found a new enterprise (e.g., in the form of a corporate entity); resources – especially capital – together with personnel and knowledge are here invested from both sides,[14] the suppliers having control over both superior resources and access to customers. For both cooperative partners, this approach is not without risks. External challengers will only be prepared to cooperate when they can protect their superior knowledge or resources. On the other hand, established suppliers will take the danger of customer abandonment into account. If cooperation between the competitors does not emerge, to the extent that a direct investment is not undertaken because of the engagement's uncertain result, a decision by the challenger against market entry is to be recommended.

The Role of a Company's Environment

The competitive capacity of a firm does not only depend on factors that it itself can determine. According to Porter very specific national conditions either strengthen or weaken its innovative capacity. The four essential conditions have been summarized in Porter's *diamond model*:

Factor conditions here denote a country's endowment with production factors. Above all so-called advanced factors, for instance knowledge, are what can form the basis for defendable competitive advantage.

Demand conditions describe the state of domestic demand for a product or service in the industrial sector. Notably, market size and the level of domestic customer aspirations influence the capacity for international competition. In a large domestic market, particularly demanding customers render an enterprise competitively capable on an international level.

Related and supportive industries are supplier or adjacent industries whose presence or absence determines a firm's competitive capacity. In this context, an important role is ascribed to the existence of company networks (clusters).

The rubric of *company strategy, structure, and competition* denotes country-specific and industry-specific management structures, work models, types of attitude, business goals, and various management ideologies. These can either promote or hinder an industry's competitive capacity. The presence of domes-

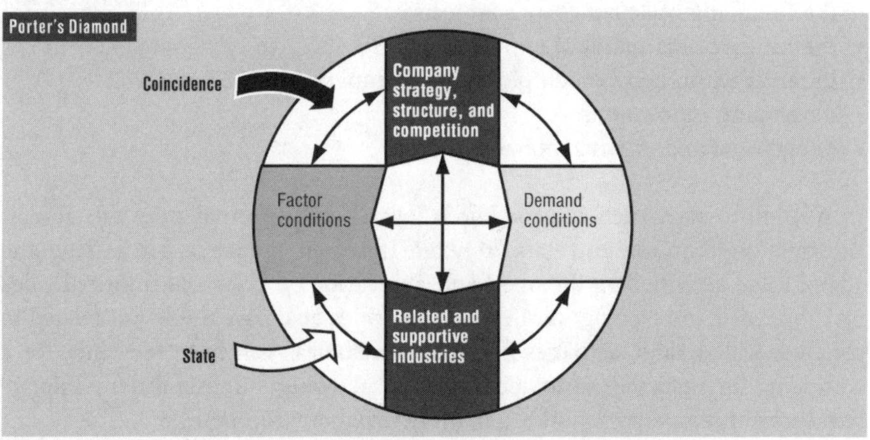

Fig. 8: Porter's Diamond[15]
Source: Porter: The Competitive Advantage of Nations

tic competition improves a firm's competitive capacity in international markets.

According to Porter, the components of international competitive capacity are not mutually isolated but rather have to support each other in order to have a positive impact. In the diamond model, two additional factors have an influence: coincidence and the state. While coincidental discoveries, crises, and other such phenomena can always have an influence on the competition process, it is above all the state that makes its mark on the diamond's positive and negative features.

By articulating both formal and informal rules, the state, together with supraregional federations such as the European Union, has considerable influence on the market process, confronting firms with two serious challenges:

- the need to analyze the extent to which their own activities and interests are affected by the *existing* rules.
- the need to analyze the extent to which their operations can be affected by potential *changes* to the rules.

Strategies for both adapting to the rules and developing (or helping to develop) them are then formulated on the basis of this analysis.

In principle, information requirements can be structured according to which added-value activities are affected by which rules. When it comes to the institutional environment, the question of the rules to be applied can be raised in relation to different activity categories (examples):

- purchasing resources;
- the production process;
- the relationship with customers (guarantees, product liability, etc.);
- the use of a product by the customer and hence customer attitudes;
- the financing of activities and projects;
- the hiring or dismissal of employees;
- the interaction between employers and employees;
- developing innovations;
- operational and organizational structure.

In relation to each such category, both formal and informal rules are in play. Information requirements take concrete form in a matrix comprising the added value activities on the one hand, the various formal and informal rules on the other hand (see fig. 9). Beyond this, rules also have a role not related to concrete added value activities but that nonetheless can have relevance for a company: for instance, various approaches allowing successful daily interaction between employees working in an external environment.

It becomes clear that for internationally active firms, there is an extremely high need for information regarding the institutional environment. Practically

Determination of Information Requirements

	formal rules				informal rules		
added-value activities	national rules	EU rules	WTO rules	...	cultural rules	moral norms, industry practice, etc.	...
purchasing							
production							
sales							
personal							
financing							
...							
activity – overlapping relevance							
...							

Fig. 9: Determination of Information Requirements

all activities in which value is meant to be produced for an ultimate buyer are influenced by rules from the institutional environment.

The degree of informational requirements is expanded by the fact that, for many decisions, simply analyzing the concrete institutional environment of a certain country or region is insufficient. For a company to make good decisions – which is to say decisions leading to defendable competitive advantages – comparisons between various rule choices are often needed. Hence a firm oriented toward biotechnical or chemical research needs to investigate the rules for research and research support in various locations, in order to then decide on one location.

At the same time, companies cannot afford to reduce their analysis of the institutional environment to the present. The formal and informal rules have too great an influence on a firm's success and competitive capacity. For this reason, it is important to identify the catalysts for changes in that environment, in order to then possibly develop the firm's own activities. Identifying these catalysts can help the firm

- have an influence on existing rules and possibly induce changes;
- recognize a process of rule-change in a timely fashion and either adapt the company's approach to the new conditions or help develop the process.

Three catalysts here take center stage: the *themes* a society is concerned with and that can be reflected in an institutional environment; *interest groups* and *interested subjects* – those affected by the themes; and the *decision process* through which themes develop into institutions.[16] In many cases, problems that have entered a public opinion-making process are resolved in the form of ordinances, laws, and guidelines. But alongside them are rules that have not been implemented in an institutionally formal way. Among these are, for example, certain codices that branches assign themselves, or else voluntary company goals or self-limitations. Even privately instituted arrangements for arbitration and corresponding procedural rules can lead to decisions over a theme.

Themes

Themes are circumstances affecting individuals in a society and in relation to which a need for decision, action, or regulation exists. Themes can be complex in nature. Identifying present and future themes here emerges as a very difficult process. In order to punctually recognize a new Zeitgeist, highly sensitive and far-reaching antennas are needed. It is also the case that such changes in cultural rules sometimes unfold over very long time spans, so that firms themselves have time to adapt their basic approach to the new rules. They receive support here from "trend scouts," institutions devoted to mapping future changes, and so forth.

On the other hand there are themes that develop within a society and that very quickly gain relevance for a firm's activities, despite initially having nothing to do with its direct business. In the course of time, these themes can themselves develop into highly concrete rules helping to determine a firm's competitive success or failure. The automobile industry, for example, is affected by many problems: recycling of materials, safety, recall episodes, limited oil reserves, the greenhouse effect, site competition. It is clear that a list of such themes – either already topical or that may well become so – could be drawn up for every firm and every branch. The range of potential catalysts behind the change of themes is very broad, extending for example from political changes (through elections, new ideas, etc.); technological progress (e.g., a lowering of communication and transaction costs); cultural changes (e.g., regarding the significance of environmental protection or the approach to health risks). The fact that such factors lead to changes in the social themes is tied above all to their touching on or altering human interests.

Interest Groups and Interested Subjects

Treatment of each of the perceived themes is strongly stamped by the interest in it by the participating parties. Depending on the theme, both they and their interests vary sharply from branch to branch and from firm to firm. But they do share one characteristic: the affected interest groups demand a regulation of themes in harmony with their interests. Interest groups are defined by a common interest (material or ideal) of the group members, with a fixed organizational structure not being absolutely necessary.[17]

In order to realize their demands, interest groups use instruments such as lobbying and other forms of political influence. The intensity of the influence here depends on the consequences that deciding on a theme will have for one or several players.

But it is not only important for these groups to know and represent their own interests; at the same time it is important for them to assess the extent to which other interest groups represent their particular – perhaps divergent – interests. What is at stake in this respect are both organized and non-organized interests. For example, in the ongoing debate over the danger and control of fine particles, there are many interest groups playing a role for firms in the auto industry. The following groups can serve as examples; each is distinguished by a particular view of a theme that can be traced back to the interests of the individuals organized within it.

Organized Interest Groups

The AAA
The AFL-CIO
The petroleum industry
Nature-protection and related groups (e.g., Greenpeace)
Political parties
Automobile manufacturers

Non-organized Interest Groups

Consumers
Citizens
Taxpayers

The Decision Process and its Potential Results

When a theme develops and sufficient pressure exists from interest groups towards deciding on the theme, then this decision will result in attitude-governing rules. This process can affect legislative authorities (e.g., the German Bundestag or the European Parliament), national and cross-border judicial authorities, and the executive, along with international organizations such as the EU and World Trade Organization.

Each organization participating in deciding on themes has its own rules. Hence for passing laws in Germany there is a system of interplay between the Bundestag and Bundesrat paving the way from opening initiative to final legislation. Analyzing the procedures at work here is an important task for companies – especially since the procedures offer the possibility to be heard and to take part in decisions.

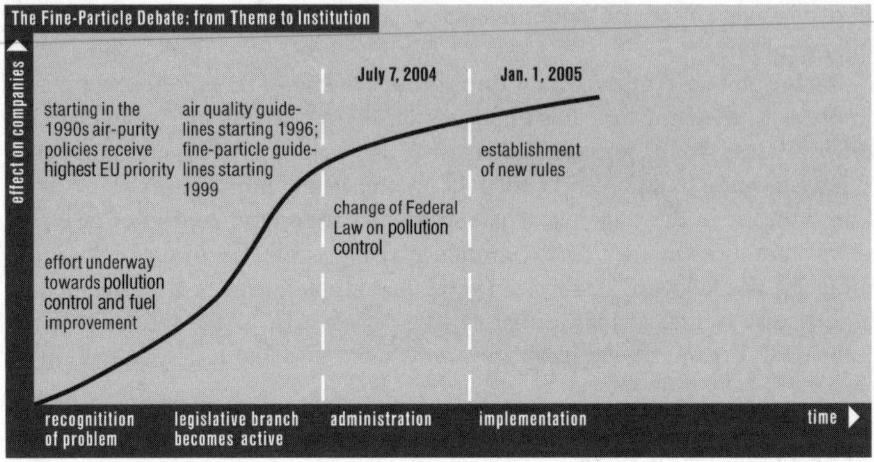

Fig. 10: The Fine-Particle Debate: from Theme to Institution

The process of a theme's concretizing into binding national rules can be clarified with a look at the political-economic debate over fine particles. The body of rules dealing with the fine-particle issue has a preliminary history within which the debate gradually emerged. Parallel to it, the fine-particle *theme* had a growing impact on relevant firms in the auto industry. Fig. 10 shows how the theme gradually emerged as an influential factor within the industry.[18]

During the 1990s, the theme of air purity gradually gained importance in public discussion. As a result of the discussion, the EU's legislative branch took up the theme at the end of the decade and issued appropriate guidelines. This created pressure on member states of the EU to pass national legislation. In 2004 relevant legislation was initiated in Germany; this became law in January 2005.

A company has numerous options regarding how to proceed during such a decision process. These range from passive waiting and then adaptation to the new institutions, to anticipatory observation of changes in the institutional environment including a policy of helping to shape these changes. Notably, a passive observation of the changes can very much be accompanied by an active adaptation to the new rules.

In this regard, what strategy promises success cannot be determined in blanket fashion, but only through a cost-benefit comparison of the various approaches. It is thus necessary to analyze the particular costs and benefits attached to the individual options.

In relation to the potential benefits of a decision, the first question to emerge involves the consequences of a given set of circumstances and the possibility of changing them. For international companies the consequences that are most relevant have to do with influence on sales figures and revenues. Such companies thus need to estimate changes in these performance indicators resulting from the institutional changes. Likewise, they will need to calculate the extent to which adaptation to the new rules will lead to costs and to changes in profit.

When the choice is made to directly influence the rules, then an impact on profits can once again be expected. For example, a firm that finds itself subject to a specific tax burden can apply for relief. The costs and benefits of such influence need to be set against those of simply adapting to the existing rules. Such adaptation could, for instance, involve a shift of production to a country with a smaller tax burden.

In regards to regulating fine particles, the rule-changes to be expected became clear at a relatively early point, companies thus knowing under which rules the competition would be performing; all competitors had roughly the same time to adjust to the new rules. While some manufacturers began the mass production of soot-particle filters for diesel vehicles, others counted on the maximum limits not being reached in the European countries. Meanwhile this assessment has been proven erroneous; the competitive position of these firms has thus been negatively affected.

Responsibility in Global Competition

Strong competitive pressure has forced firms to exploit all maneuvering room for increasing effectiveness and efficiency. Within this process, their activities can unfold either according to the extant rules or with the aim of influencing the rules. In either case, conflicts can arise between economic rationality and moral behavior. Competitive capacity might thus be intensified through morally unacceptable measures such as the use of child labor – even though in certain countries such measures remain legal. By *morality* we understand the practically lived arrangement of norms and values. *Ethics* is concerned with the theoretical grounding of these norms and values; it places the reflective process above practical morals. As a branch of philosophy it studies the motives, forms, and consequences of human action and the question of what defines good and bad actions. *Economic ethics* has the goal of uniting rational economic action and ethical reason.

Karl Homann has closely considered the moral dilemmas in which many managers find themselves. In this context, he has proposed a four-quadrant scheme (see fig. 11). Within this scheme, the two axes describe the effects of a specific company action in relation to moral acceptance on the one hand, and profitability of the company on the other.

Quadrant 1 refers to a complementary relationship between moral behavior and economic success. This can be grounded in the economy's overall regulatory framework demanding and effectuating moral behavior from all competitors to the same degree, so that with observance of the rules economic damage through sanctions is absent. Or it can be grounded in a profitable "sale" of moral behavior, for instance through its effective use in publicity.

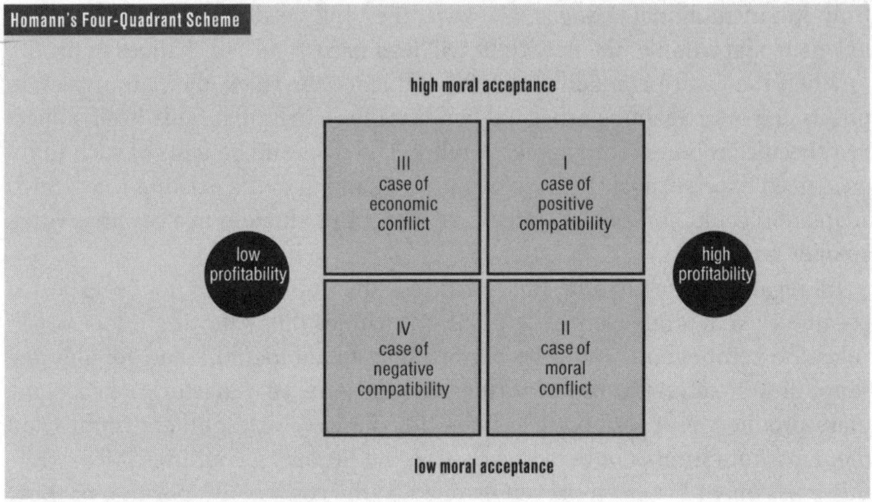

Fig. 11: Homann's Four-Quadrant Scheme[19]
Source: Homann: Marktwirtschaft und Unternehmensethik

Quadrant II refers to the case of a company's behavior not being in harmony with a society's moral ideas, even if it is legal. The company here opts for profitability and against morality. It mainly justifies this decision – to the extent that it is at all justified – with references to competitive pressure. As a response to the question of an action's moral legitimacy, phrases such as "right now we really have other problems" are symptomatic for this situation.

Like quadrant II, quadrant III refers to a conflictual relationship between profitability and moral acceptance. The basic difference is that in this case the firm has opted for morality and against profitability. Quadrant IV refers to an action that is neither morally accepted nor profitable. In this case a decision to exit the market can be expected.[20]

Homann's schema makes clear that moral behavior can be highly problematic for a company in conditions of strong competitive pressure. In this light, it is crucial to note that, in principle, a general regulatory framework consistently integrating and enforcing moral expectations can effectively counter conflicts between morality and economic rationality. Moral expectations would here be fixed as a restriction on all market participants; competition would take place in terms of moving according to the rules of the game, not through violations of moral norms. Homann's approach, inserting economic morality into the framework of general economic regulations, frees the firm from permanent justificatory pressure. Actions consistent with the regulatory framework can be considered ethical – this of course to the extent that the framework is itself in harmony with a society's moral concepts. Upon such a basis, a firm can focus on purely economic competition.

It needs to be stressed that the framework's existence does not free a firm from moral responsibility. A purely profit-oriented approach is only justified when the general conditions do not reveal an ethical deficit. But this can hardly be assumed: the various national governments have scant possibilities to impose an effective regulatory economic order on internationally operating companies. This deficit is only partially overcome by rules formulated on the level of regional organizations and the WTO and monitoring by NGOs (Non Governmental Organization). In reality then, the conflictual situation defined by cases II and III will often prevail. As a reaction to this by the companies, Homann considers two types of strategy especially practicable: a strategy centered on competition and a strategy centered on considerations of political order.

In the competition-centered strategy, moral behavior is aimed at because those engaged in the transaction have had or develop a preference for partners with moral integrity. This preference can apply to highly varied objects: products, production, orientation toward employees, orientation toward the environment, and so forth. What is here essentially at stake is discovering profit possibilities that can be grounded in morally acceptable policies and actions. In this manner, moral and economic goals become mutually aligned.[21]

Concepts such as corporate social responsibility (CSR) should themselves be understood as reflections of competitive-centered strategy. CSR signifies ethical company behavior and, beyond the standard entrepreneurial activities, exercise of good citizenship, which is to say acceptance of communal responsibility.[22] However, implementation of such a concept can only be counted on when it is tied to economic success for the company: "No one can expect a company to accept serious economic disadvantages on grounds of moral behavior while less moral competitors pocket the profits."[23]

Even when a competition-centered strategy is not feasible, firms are not exempt from their ethical responsibility. In that case, they need to use the strategy centered on political order, referring here to the deficits in the existing regulatory framework and working on their reduction.[24] The ongoing active support for the Global Compact initiative of U.N. General Secretary Kofi Annan is an example of this alternate strategy. The initiative was presented at the World Economic Forum in Davos in 1999. Its goal is to strengthen cooperation between the United Nations, companies, and social organizations and make it useful for realizing essential U.N. objectives. To participate in the Global Compact program, companies and organizations commit themselves to act according to ten principles pertaining to human rights, labor relations, the environment, and anti-corruption. The principles, which are derived from central United Nations goals, are listed in fig. 12.

The participating firms report regularly on their progress in regards to the various realms covered in the initiative; they also publish a homepage (www.unglobalcompact.org), in order to serve as examples for other firms and to offer both NGOs and the public a chance to comment. Again, it should be stressed that if the Global Compact's rules are not generally implemented, their acceptance can itself lead to competitive disadvantages.

Principles of the Global Compact

Human Rights	
Principle 1:	The support and respect of the protection of international human rights
Principle 2:	The refusal to participate or condone human rights abuses
Labor	
Principle 3:	The support of freedom of association and the recognition of the right to collective bargaining
Principle 4:	The abolition of compulsory labor
Principle 5:	The abolition of child labor
Principle 6:	The elimination of discrimination in employment and occupation
Environment	
Principle 7:	The implementation of a precautionary and effective program to environmental issues
Principle 8:	Initiatives that demonstrate environmental responsibility
Principle 9:	The promotion of the diffusion of environmentally friendly technologies
Anti-Corruption	
Principle 10:	The promotion and adoption of initiatives to counter all forms of corruption, including extortion and bribery

Fig. 12: Principles of the Global Compact[25]
Source: United Nations, 2005

For this reason, the strategy centered on political order does not eliminate the possibility that despite actively promoting correctives to the general regulatory framework, in competition companies will display behavior that is not morally accepted. At first sight, this behavior may seem to reflect a double-sided morality. Frequently, exigencies of the competitive market process make political self-contradiction a company's only option.

Conclusion

Globalization and the disappearance of trade barriers will continue to strengthen market forces. Effective and efficient companies are rewarded with market share and profits. Inefficient companies will not be able to hold their ground competitively; they will not be able to gain any new markets and will lose those they have.

Consistent market orientation, in the sense of orientation towards both customer needs and the competition and a corresponding structuring of the entire added value chain, forms the guideline for management in global markets.

In the creation of competitive advantages, the business environment plays an essential role. National characteristics and supra-national rules (formal or informal) influence the innovative strength of companies and have an impact on the outcome of the market process.

International firms must be aware of their responsibilities towards human beings. In the long term, an exclusive orientation towards shareholder value places the free-market economic system in question.

2
From Selling Goods to Marketing Services

Olaf Plötner

The Significance of the Services Industry in B2B Markets

"That was no evolution – that was a revolution," recalls an IBM manager who witnessed Lou Gerstner's transformation of IBM in the mid-1990s. The CEO arrived at that firm in 1993. Instead of continuing to focus on the manufacture of high-quality computers, he decided to press ahead as energetically as possible in promoting the company's servicing business. This decision meant the dismissal of 36,000 employees who, it was felt, were incapable of accompanying the company on its new course. The CEO who then took over at IBM in 1999, Ron Palmisano, continued to press ahead with this services-oriented strategy. At present more than half the company's turnover of 47 billion dollars comes from the services sector, which is thus responsible for a significant portion of the company's profits.

But why did IBM undertake this radical change? From a short-term perspective, it may well have been the operative pressure to bring the company out of a loss-making zone and save it from threatened divestiture. But long-term strategic considerations also favored the new orientation, which was based on a recognition that even goods of a presumably sophisticated technical nature such as computers could be copied with ever more speed by less expensive firms. At the same time, it was grounded in a hope that at least for services with a high degree of complexity, this might not be the case to the same degree. People were convinced it would be possible to gain a lasting advantage over international competitors on the basis of the special qualifications of IBM employees in the new areas of business.

The assumption is in fact solid that fewer employees with especially high qualifications take part in producing IT hardware than in providing complex IT services. The decisive factor here is not only employees' technical competence, but above all their social competence. In the IT services business they have more contact with customers, have to demonstrate a higher degree of intellectual flexibility, and must think more holistically. Importantly, when it comes to such competence many firms in western Europe and North America believe they will be able to maintain a lead over China or other "emerging countries" for some time to come.

This in any case only holds true for services in B2B markets that demonstrate a relatively high degree of complexity or difficulty. Hence this discussion

Bringing Technology to Market. Olaf Plötner and Robert E. Spekman

ISBN 978-3-527-50270-7

will focus on only these services: for instance, the customer-specific architecture of a telecommunication network, project management in the construction of a chip factory, operating a complex IT-system, inspecting and overhauling an airplane turbine, and so forth.

Researchers have offered a number of definitions meant to distinguish the sale of goods from the sale of services.[26] The following discussion will take a much-simplified approach to the distinction, the sale of services here being understood as a pendant to the sale of goods, with the central difference involving contact with the goods. Hence when goods are being sold the customer can physically handle what is being purchased, while when services are being sold this is precisely not the case.

At the same time, these two forms of business are closely connected in many ways. For example, when a customer acquires an industrial turbine – in our context, a transaction involving goods – this is usually tied to the purchase of a maintenance package, understood here as a service. In actuality, for providers the marketing of goods is not possible without services being supplied as well. This becomes even clearer when we note that transporting goods to a customer already represents a service. To this extent the "pure" sales of goods and the "pure" sales of services simply serve as polarized categories between which different firms can be classified. To be sure, when it comes to marketing questions the correct classification is very important; for – as will be shown – there are important differences depending on whether essentially goods-based or essentially service-based offerings are going to be marketed.

A pragmatic distinction between sales of goods and sales of services is also commonplace within companies. In this regard, many B2B companies are downright fascinated by the high profitability of domains they subsume under a "service business" rubric. The high economic attraction of services is underscored in fig. 1, which presents the results of a study comparing the profitability of goods-oriented and service-oriented business in selected B2B branches. It

Profitability Comparison of Goods and Services Business

Margin differences of OEM- and services business

Industry Examples	Margin in OEM-Business	Margin in Services
Power Plants	2–5%	15–20%
Locomotives	3–6%	8–10%
Machine Tools	1–12%	5–15%
Paper Machines	1–3%	10–15%
Metallurgical Plants	-3–+6%	15–20%

Fig. 1: Profitability Comparison of Sales of Goods and Services[27]
Source: Monitor Study, Industrial Services Strategies

would appear from these results that in their strategic decisions many firms feel motivated to expand their service activity.

The separation of goods and services according to marketing logic can take very different form, depending on the range of services offered. The narrowest connection emerges when the services marketed by a firm are directly tied to the use of goods it is itself supplying. This is the case, for instance, with the marketing of one's own machines. An expansion of this service offering takes place when competitors' products are also being maintained. Siemens Medical Solutions thus maintains not only its own radiological equipment but that of General Electric and Philips and vice versa.

As the next stage in developing their services activity, providers can offer services with no direct relation to their established products that is perceptible to customers. For example, rather than marketing turbines and connected services, a manufacturer of gas turbines might supply energy to its customers. One output-oriented pricing concept frequently agreed on in such transactions, "power by the hour," already makes clear that the input-relevant selection of goods is here of secondary importance for the customer.

Beyond this, another possibility is that a provider very consciously prefers to avoid any external connection between selling goods and selling services. This is the case with, for example, so-called competence commercialization,[28] in which a provider wishes to market competencies gained in the goods sector as consultation services separate from that sector. Lufthansa Consulting GmbH thus offers other airlines its know-how regarding route optimization, and Siemens Business Services is interested in selling strategic consultation in the telecom and IT sectors. Frequently companies endow such service branches with an independent legal form, so as to work against possible suspicions by customers that the consultative activities are aimed at supporting the sale of one's own goods.

In any event, the present euphoria shared by many Western B2B companies regarding that business needs to be considered critically. For it is by no means the case that all firms have arranged a shift from goods to services as successfully as IBM. Many have only managed to generate weak profit margins; others have entirely vanished from the market. Interestingly, degrees of success in providing services can even vary between branches of the same firm. Where, for instance, Siemens AG did not succeed as expected in the realm of IT services, it is nevertheless considered an important benchmark for all competitors through its similar activities in energy production and both medical and automation technology.

In analyzing the sources of success, one thing is repeatedly clear: it is not enough to simply supply more services to the customers or increase their range. This is underscored by the results of one study of the German machine-construction industry. This study is not completely up to date and was limited to German firms; but the problems it points to for B2B firms still exist in many countries.

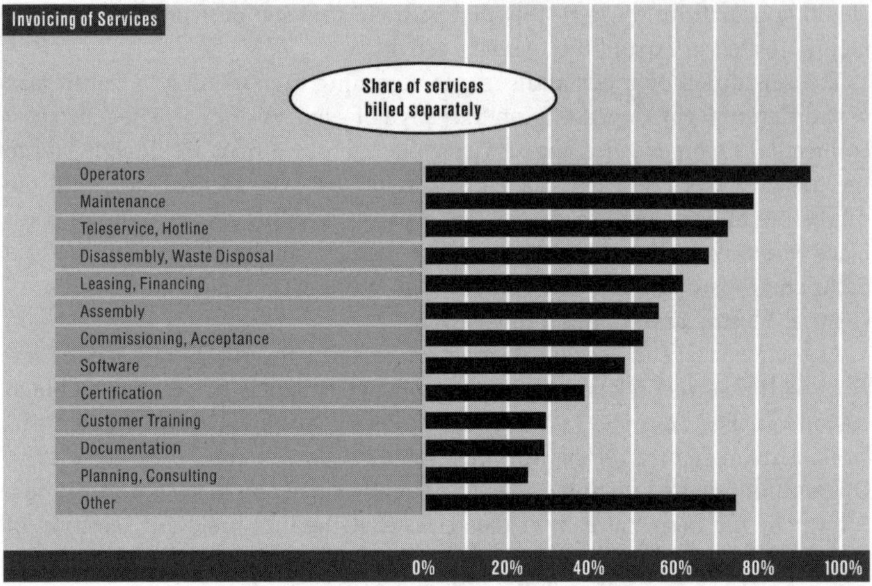

Fig. 2: Invoicing of Services[29]
Source: VDMA-Study, Product-Related Services

First of all, these results confirm the correctness of the intention of many B2B firms to expand their service activity. But on the other hand, they make clear the great challenge accompanying this intention. It lies in being able to include suitable customer charges for services rendered. Put pointedly: services are very profitable – but only when customers are persuaded to pay for them.

It is striking that for firms whose roots lie in the sales of goods, this problem emerges above all in relation to services with a high growth prognosis, such as consulting. Precisely when such services are made use of by customers before a purchasing decision involving a large plant or machine, there is very little willingness to pay. Notably, at present this is even more the case for the Western European than for the North American market. Many customers define consultative services simply as part of a provider's acquisition activities, no matter how costly they may be. And in the end, it is customary to compare only the price of the goods in the offers for the plant or machine; the costs of the provider tied to the consultation are not at issue.

From fear of not achieving their ambitious sales goals, many sales managers accede to such an understanding by the customer. Many are not even aware of the value of the services that have been given away. Others hope to somehow cover the costs of the unpaid services through the sale of their product – when precisely in machine construction the high pricing pressure often reduces such hopes to nothing. This approach in terms of putative cross-subsidization is especially questionable when a provider, fully following the trend of the times, steadily increases his resources for performing services accompanying the sale of goods. The rising costs for personnel in this realm then have to be covered by ever higher prices for goods, which, however, will be accepted by the customer with ever more difficulty.

For this reason it seems sensible to consider the customer value of *all* services performed, and to selectively accelerate their professional marketing on that basis. But often just those companies with roots in the sale of goods have difficulties in successfully marketing services as soon as they go beyond the normal maintenance and repair business. These firms quickly notice that the rules and experiences involved in selling goods cannot simply be transferred. Rather, the previous approaches have to be called into question and new marketing paths discovered. The following pages are meant to offer suggestions regarding possible directions for such paths.

Specifics of the Services Industry

Before offering recommendations regarding company policies, it seems appropriate to more precisely consider the basic differences between marketing goods and marketing services. By way of example, two marketing processes from the realm of consumer goods will be considered here: selling goods in the form of a fountain pen and selling services in the form of a haircut. An assumption here is that the two transactions have comparable monetary value. The examples are meant to be understood as representing a goods-services polarity, in full awareness that in the B2B practice at issue here, these types of business are rarely found with such distinct characteristics. Rather, we here find mixed forms, which is to say that transactions have both material and immaterial dimensions. To this extent the two examples are not directly transferable to complex businesses in the B2B area. Nevertheless, they are highly useful in illuminating basic differences in patterns of sales and purchase – differences from which we can derive important insights for the complex marketing of goods and services in the B2B markets.

If we wish to purchase a half-way high-quality fountain pen, we normally go to a stationary store or corresponding section of a department store. We find there a wide range of models displayed behind glass; these have been manufactured by firms from all over the world. We consider the products and have the salesperson hand us those we find optically attractive, with an acceptable price. We test their "feel" and use a writing pad to see how each pen writes. Finally we choose the pen we like most, go to the register, and pay for the selected product.

When we wish to have a haircut, we usually go to a barber or hairdresser we know or at least one who has been recommended by a friend. We enter the barbershop or salon, take an assigned seat, and discuss with the haircutter how our more or less luxuriant hair should look at the end. The haircutter then begins to work, now and then asking us to tilt our heads in one or another direction, this request often taking the form of a marginal remark in a broader

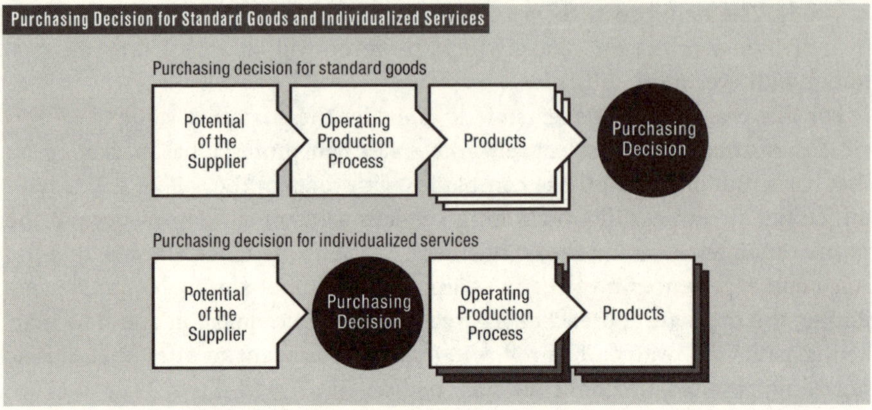

Fig. 3: Purchasing Decision for Standard Goods and Individualized Services

conversation about indifferent or significant matters of daily life. In the end, we are asked to scrutinize the previous minutes' results in a mirror. Most of us will then express our satisfaction, go to the register, and pay for the service we have received.

In structurally comparing these two transactions, a simple model may be used to distinguish the *potential* to produce goods and services, the *process* of producing them, and the *results* of the process. The first category takes in the resources that a provider wishes to use in creating value, for example machines, materials, or employees. Following preparation of these resources, the process of providing the goods or services is initiated. Its results are, for example, the manufactured fountain pen or the paid-for haircut.

A first basic difference between transactions involving fountain pens and those involving haircuts becomes clear in relation to the question of time of purchase. In the above model, we see that in the pen transaction the production results are already present before the purchasing decision; the product has already been made somewhere before the customer knows he would buy it. The situation is different in the case of the haircut; here the buying decision takes place before the service-provision process is initiated. Fig. 3 clarifies this distinction.

The next difference emerges when we look at the execution of the process through which the goods or services are created. The above-mentioned pen is produced without the customer being involved in the process. In the case of haircuts, customers are involved – indeed actively so. They express their desires regarding the way the hairdo or haircut should look – that is, regarding the results. They hold their heads still or tilt them, according to the hairdresser's suggestion. Three additional important insights regarding the servicing business follow from this example. First, customers are involved in the service-provision process in a basic way. They are, so to speak, "co-producers" of the services they purchase. Second, the common production work of provider and consumer always makes interaction between them necessary. And third, because of the customer's integration into the service-provision process, such services are always customer-specific. The results at work here are thus individualized in a basic manner.

The last characteristic of the classical service business to be discussed here emerges from the above-mentioned points. If customers participate in the service-provision process and the purchasing decision takes place before emergence of the process's results, it is only logical that they cannot test beforehand what they will receive later. With a pen, it is possible to judge whether its design is pleasing, it can be gripped comfortably, writes as one would like, and so forth. In contrast, when customers visit a hairdresser they cannot know how the haircut will really look before making the purchasing decision. In other words, it is impossible to test the quality of the results before deciding to purchase a service.

Since this facet of the service business has special significance for its marketing process, it merits a more detailed description in our context. The first thing we should note here is that no matter which type of business is in play, for customers every purchasing decision is bound up with risk. They are consistently faced with the question of whether what they are acquiring is in the end worth its price. The question arises because at the time of the purchase decision, much of what they expect to be the benefits of the product or services purchased still lies in the future. In this way a consumer cannot yet know if the wine he or she is buying today will taste of cork tomorrow. Similarly, a delivery firm must be in the dark over whether its newly purchased semi tractor trailer will actually display the expected longevity in the coming years. Since the future can never be predicted with certainty, customers feel insecure.

Normally customers wish to diminish this sense of uncertainty. For this reason, they seek information abut the quality of the goods or services that may be purchased. Hence the delivery firm will obtain information on the quality of the truck's engine, ask other customers about their experience with this particular model, have an experienced driver take a test drive. Of course considered objectively, information collected in this way cannot generate any complete certainty concerning the truck's future reliability; but it can at least significantly reduce the uncertainty, to the point that customers are ready to "talk business." Correspondingly, providers will do their best to furnish customers with information tending to diminish the uncertainty, which is to say anxiety at making a false decision. Specialists define this process of information searching and information transmitting between consumer and provider in the framework of a purchasing decision as screening and signaling.

Information transmitted to reduce this uncertainty can be related to each of the above-mentioned goods-and-services-related categories of potential, process, and results. Admittedly, customers do not have access to all information on a product or service before the sale. Rather, they have access to information related to so-called search qualities: the design of a fountain pen, easily assessed at that point by the customer, can here serve as an example.

Other qualities can, to the contrary, only be assessed after the purchase, for example whether a truck will chalk up the hoped-for mileage. Naturally, a prospective customer can sometimes draw on experience regarding the mileage of similar models. But an evaluation of the quality of the specific vehicle being eyed for purchase will only be possible after experiencing years of driving that very vehicle. Specialists thus here refer to "experience qualities."

Finally, there are also qualities that cannot even be meaningfully assessed by a customer after the purchase. This is the case, for example, with the quality of recommendations regarding strategy that a consultant gives a client company. When the company then experiences economic success, this could have resulted from having received good strategic advice, but equally from the emergence of favorable factors in the national or international economy. Similarly, a lack of economic success could have been the result of bad strategic advice, but also of the flawed implementation of advice that was actually solid. Hence in the end, customers cannot conclusively confirm that the dearly purchased counsel was worth the money. Rather, they must put trust in its quality. Specialists here speak of "credence qualities."[30]

Against this background, we can now further consider the fact that the quality of a service cannot be tested before a purchasing decision is reached. Using the above-cited terminology, this means that the results of the service provision process have no search qualities whose scrutiny might help customers reduce their uncertainty before the purchase. When it comes to experience and credence qualities, the possibility of assessing these results is absent by definition.

For this reason, in such transactions customers reduce their sense of uncertainty through something like detours, two approaches here being especially noteworthy. In the first place, there is increasing recourse to personal experiences, or the experiences of other customers, with providers of comparable services. Hence the quality of the purchasable service is deduced from the experience qualities of similar services. In the second place, customers inform

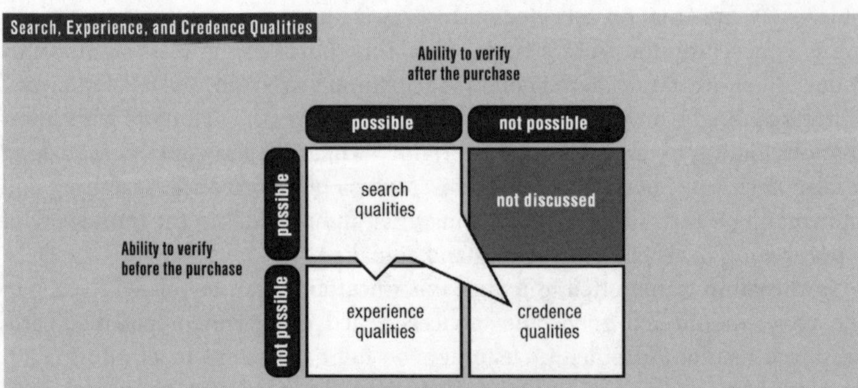

Fig. 4: Search, Experience, and Credence Qualities[31]
Source: Plötner. Das Vertrauen des Kunden

themselves to a greater degree about the service provision process and provider's potential in this respect in order to draw conclusions about credence qualities. Expressed simply, the customer reflects as follows: If the provider seems competent and trustworthy to me, then the results will be good; on the other hand, if I'm not convinced of the quality of the provider's potential, the same applies to the results he will deliver.

These customer's screening approaches can of course also play a role in the sale of goods. Here as well, customers can reduce their uncertainty by informing themselves about the seller's potential. But notably, he usually obtains such information simply to supplement that about a product's search qualities. Since just these are missing in the service business, customers have other focal points for reducing their uncertainty. We can thus summarize as follows: experience and credence qualities have a higher significance in the sales of services than in the sales of goods.

Understanding this point is particularly important for providers wishing to reduce the scope of their goods-related business and increase the scope of their services-related business. The point already makes clear that changes in the portfolio of what a firm offers force an entire series of changes in other marketing-relevant company areas. Communication policies have to be altered, along with sales organization, customer management, personnel policies, and a great deal more. The following section will discuss some concrete approaches for possible changes. It should be emphasized that these are not meant as a blueprint for unreflective acceptance but simply as an indication of the directions in which a company might best steer its transformation processes.

Implications for Sales and Marketing

When it comes to the customer-related communication policies of B2B firms shifting their emphasis from goods to services, the above discussion suggests a number of measures. In the first place, we should note that as indicated, no results of the service provision process allowing a provider to reduce the customer's perceived risk can be available before the service's purchase. Hence by their very nature, search qualities cannot be presented in the framework of policy-determined measures designed to remove customer doubts *before* the purchase decision.

Customers have to be convinced of the experience and credence qualities of the service. These can themselves not be established on the basis of the specific service in which a prospective customer is interested. For experience qualities, the communicative use of references instead takes on salient importance. The service is thus promoted through the good experiences other customers have had with similar services from the provider. The more similar the service and credible the referring customer, the better the communicative effect on the target customer.

For credence qualities, which in principle are removed from any result-related control, the provider's communication will not refer to the service itself. Instead, information can be given the customer concerning the potential of the provider firm, for example the qualifications of its employees. Professional consulting firms thus normally inform their customers of the career histories of the consultants meant to take part in a prospective project. The intention here is to convince the customer of the employee's competence and thus gain trust in the high quality of the work to be offered.

For many providers of industrial services, such a pronouncedly employee-oriented communicative structure is not yet standard. But in fact, there are few if any reasons why informing operators of, say, technically complicated machines about the competence of those to be responsible for repairing and maintaining the expensive equipment would be anything other than a confidence-building measure – one of great benefit to the equipment's provider.

Over recent years, certification has been increasingly used as a tool of communication policy – for instance by indicating that a given service conforms to the EN ISO 9000 series of quality-management norms. In this way the customer receives information about the capabilities of the service provision process. The information is meant to demonstrate that the work is careful and well organized, and that there is corresponding process management know-how. It is the case that in this approach the employees are not directly thematized; but in terms of substance, here as well we find a positive description of a firm's potential to provide services. And here as well, the description is meant to reduce the customer firm's sense of risk and convey that with this provider, it is in good hands.

Naturally, for service-oriented firms there are a number of communicative options in addition to supplying information about service provision potential. Especially noteworthy in this context are the possibilities referred to in the specialized literature under the rubrics of branding and corporate identity. It is the case that in view of the limited possibilities for communicating about specific results in the service-oriented business, creating a general company image has relatively higher priority than in businesses focused on goods.

At the same time, in the service business it is especially important not to reduce communication policies to advertising or PR. Rather, a more comprehensive approach has to be developed – one taking in all phases of sales and after-sales activities. We can here simply note that the integration of customers into the service provision process so necessary in the services sector poses special communicative challenges for the provider's employees. They have to become customer advisers, and they need to be able to work with customers in an intense and convivial manner – without, to be sure, losing sight of the firm's economic interests.

Precisely in firms in the process of shifting from goods to services, employees often lack these communicative skills. For this reason, special demands often emerge here for personnel management – most so in relation to proper choice of personnel, but also requiring further training. Experience tells us that special efforts are essential not only to impart the necessary skills but even simply to create an openness for the necessary learning process on the part of employees. In this regard, it cannot be assumed that the efforts by the employees always culminate in the desired success: one important reason for the above-mentioned development at IBM being accompanied by a dramatic change of employees.

The strong role that credence and experience qualities play in complex services is not only decisive for a provider's communication policies. At the same time, it has important consequences for policies regarding pricing and conditions. In classical goods-centered businesses, performance can be more clearly specified and results more easily reviewed. This is the basis for agreeing on prices for the results and establishing clear regulations for dealing with shortcomings. But what specifications are suitable for a consultant's strategic project? How effective can warranty provisions be if performance quality cannot be validly measured? And why is the per diem rate of one consulting firm actually twice that of another such firm?

The non-verifiability of the service provision results before and even after the purchase leads to agreements that can be markedly different from those at work in the classical approaches to fees of goods-oriented businesses. One approach that has become particularly common over recent years is to set prices not for the actual performance result but for what is supposed to be created through this result. IT-service companies thus no longer sell their repair and maintenance services on the basis of working hours but on that of the network's subsequent availability.

With such developments, providers of complex services are often following market pressures; but the developments can be tied to risks that are not negligible. In the first place, these providers cannot be the sole parties influencing the quality of results of the service provision process. Rather, they are dependent on the contributions made by customers during the process. If these contributions do not meet the providers' requirements, this will have an effect on the entire results and the provider firm runs the risk of not being suitably paid for its own service outlays. The risk is all the greater when providers tie their prices to the client company's economic success. Their influence on such success is usually very limited. Although customers are here offered the great advantage of their costs being variable, providers may not be able to cover their own costs if the client company makes marketing errors.

In general, close and often convivial relations between provider and customer emerge in business with complex services. Nevertheless, a provider will only push through such pricing models if this is tied to the chance – very rarely at present in industrial enterprises – for especially high returns, or when the competition situation absolutely demands it. If this is not the case, there will be an effort to market the services without taking commercial risks – for instance to charge on the basis of employees' work time. To be sure, high daily or hourly rates presuppose complete trust by the customer in the provider's skills and efforts.

For building such trust, it is very helpful if the person in the provider-firm who has promised something to the customer is the same person responsible for eventually providing that service. This represents a basic difference with classical goods-centered business, in which production and marketing are rarely in the same hands. This is acceptable for clients since they can reduce their uncertainty through, for example, the product-control process. With complex services that possibility has to be replaced through trust in the provider or its employees.

This has important implications for the provider's sales organization. We can here return to the example of consulting firms such as McKinsey or Boston Consulting: how large is the sales division of such firms? The answer is: there is no sales division. In fact the firms' partners are responsible for both the sale of the projects and their subsequent successful realization. This means that after the customer has confided in one of the partners – i.e. explained the company problems to be hopefully solved through the purchase of external support – the partner is available for the project's duration as the responsible contact person. In this way consulting firms have no function-specific organizational forms providing for heads of sales and production. Rather, the structures best suited for providers of complex services are customer-oriented and project-oriented.

In this respect it is characteristic of consulting firms to have partners – in other words employees on the highest step of the hierarchy – responsible for the sales function. Typically, the competence for taking over project responsibility has accumulated in the course of their consulting careers, involving years of work in the service-provision process. Only those who were most successful in mastering these years will have a chance to profile themselves in that function. With some exaggeration, we can say that only the best people in production advance to sales agent positions, and it is self-evident that reaching this career goal is tied to a significant increase in the already high salary received at a consulting firm.

This approach is at least untypical of the company structure of classical goods-centered businesses. It is the case that over recent years the status of sales managers in these businesses may have improved. Nevertheless, now as before, in the technically oriented B2B firms it is customary for those with the highest rank and respect to be located in research and development rather than in sales.

It is just as rare to find structures established in which the responsibility for production, sales, and even subsequent customer service are assigned to a single position – let alone that of board chairman or managing director. Rather, various divisions are responsible for these activities; although their work is meant to be coordinated in practice, not insubstantial interface problems arise.

Undeniably, such function-related structures have often proved their value for goods-oriented firms; they can be entirely sensible and well-grounded. Nevertheless, the more companies have developed their complex-services business, the more necessary it seems for them to call these structures into question. With such development, internal structures need to be rendered more flexible and, thinking in terms of functional division, increasingly replaced by customer-oriented project responsibility.

Conclusion

The above observations first focused on the significance of business with complex services in B2B markets and reviewed the essential market-relevant differences between the provision of goods and services. This was followed by a description of approaches to developing sales and marketing in B2B companies shifting their business orientation from goods to the marketing of complex services. In relation to this development, five important points are underscored once again below:

1. In the contents of marketing communication, both the description of service provision potential (especially on the part of the firm's own employees) and firm-related branding gain in importance.
2. The stronger integration of the customer into the service provision process requires special capacities for personal communication by committed employees; these capacities should be closely considered in personnel development.
3. If a service provider is risk-averse, the firm's pricing policies should be oriented around use of its resources, not around the service provision results. (Since both provider and customer are involved in the service-provision process, the results are not influenced by the provider alone.)
4. With complex services, responsibility for sales and production should be in the same hands (more project-specific or customer-specific instead of function-specific organizational forms).
5. This clustering of responsibility demands a high degree of competence and authority from the various employees; sales tasks thus become an executive matter.

These points are meant to simply suggest the direction of changes in sales and marketing tied to changes in what a firm offers. It is clear that one particular question, of interest to many firms, cannot be answered in this context: how strongly, for instance, communications policies and the sales organization should be altered in what stage of these changes. In this regard, there are an endless number of options whose practical realization must always also be viewed in connection with general conditions specific to a firm. It is in any case certain that other marketing-relevant areas need to adapt when a provider-firm shifts its focus from goods to services. Although this insight is banal, the adapting is either absent in practice or comes too late. For in this thematic realm as well, resistance to innovation is often greater than willingness to change. It is the job of management to mastermind suitable developmental steps and counteract any such reluctance to move forward.

3
New Value Chains

Michael Kleinaltenkamp

Developments in Consumer Goods Markets Trigger New Value Chain Structures in B2B Markets

All goods and services are produced for the purpose of consumption, but the days when people actually consumed self-produced goods are long gone. Instead modern business is characterized by a division of labor, with several market stages between original production and consumption combined with different manufacturers and producers, processors, traders, and service providers. We are thus largely dealing with markets in which goods and services are purchased because they are directly or indirectly required for the production or sale of other goods and services.[32] The buyers in these markets are consequently sellers in their own markets, and demand is always *derived*; that is, it depends on downstream buying decisions and trade stages possibly going back to the initial request of the end consumer.

What additionally characterizes our mature economies is an increasing demand for tailored consumer goods, a trend that most probably is going to continue[33] and that has led to constant product differentiation. To what degree

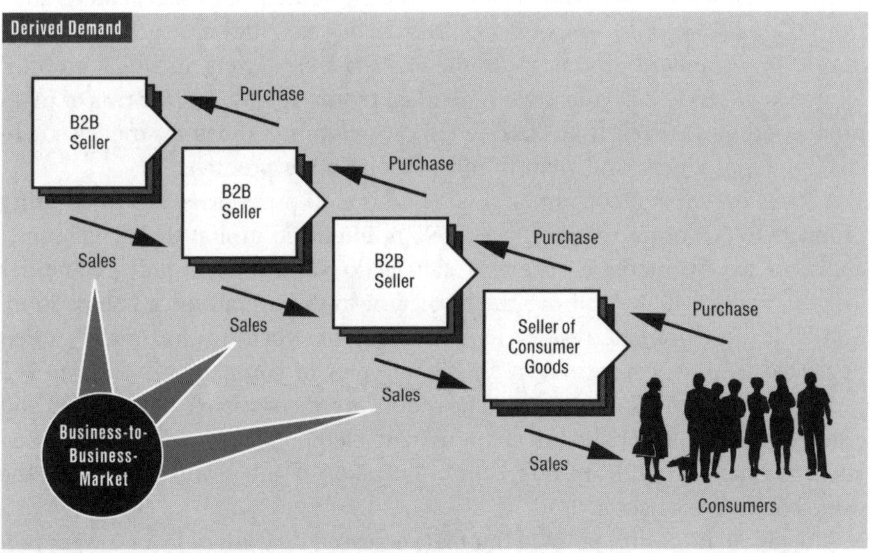

Fig. 1: Derived Demand

Bringing Technology to Market. Olaf Plötner and Robert E. Spekman

ISBN 978-3-527-50270-7

this has been caused by highly individualized consumer demands or encouraged by marketing and sales strategies is a matter of debate.[34] In either case, it has divided markets into smaller and smaller slots and resulted in a corresponding diversification of goods. Initially, this process meant offering larger numbers of models and variations, a development for which the auto industry serves as an interesting example. (Over the past thirty years, the number of auto types per model has increased by five to eight times.) Similar developments may be observed in other industries.

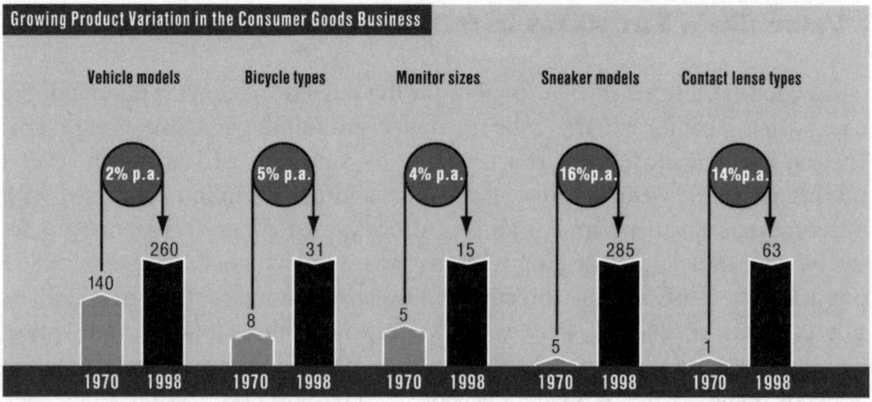

Fig. 2: Growing Product Variation in the Consumer Goods Business[35]
Source: Cox and Alm: The Right Stuff

The second trend or development goes a step further. This trend is frequently referred to as *mass customization*. This means that a large number of consumers are offered products tailored to their demands. Products then range from the *match-to-order* or *locate-to-order* variety – i.e., they assist the customer in selecting standard products – to the *make-to-order* variety, in which an offer for goods or services is tailored to individual requirements. Such forms of mass production may currently be observed in cars, clothing, shoes, cosmetics, kitchens, media products, and vitamin pills, among other products.[36]

A focus on tailored consumer goods is above all profit-oriented. In tailoring products to customer requirements, sellers intend to exploit their customers' ability to pay, to increase customer satisfaction and thus improve customer lock-on, and finally to improve their position in the overall price competition. Still, this development results in non-negligible costs (unfortunately often neglected in practice), primarily in higher costs of complexity.[38] Customized products and services require that customers' needs or wishes be specified and entered into the seller's value chain as new elements.[39] As will be discussed later, this has a significant effect on the division of labor and thereby on the value chains in question.

Already Adam Smith pointed out that increased division of labor lowers production costs while increasing output. As it happens, this increase also

Types of Customization

Form of customer integration	Interaction point	Degree of customer integration
Match-to-order/locate-to-order Assistance in selection of existing standard products	Trade/Sales	Soft customization
Bundle-to-order Bundling of standard products and services to a customized product	Trade/Sales	
Assemble-to-order Customized final assembly using standardized components and modules	Final Assembly	Hard customization
Make-to-order Customized production including customer-specific components	Production	

Numbers of customer-specific activities

Fig. 3: Types of Customization[37]
Source: Piller et al: Industrialisierung

requires a higher degree of coordination and cooperation among the different performers, be they individual workers, corporate divisions, or entire organizations, resulting in an increase of overall costs. The more specialized the actual processes and process operators, the higher the resulting costs of coordination.[40] This can be seen in fig. 4: while production costs fall with the increased division of labor, costs of coordination grow. The highest point of cost efficiency in the labor division (D^{opt}) is therefore located at the point where the mathematical curve (total costs of production and coordination) reaches its minimum (C^{min}).

But if costs of coordination actually do increase with higher customization, the cost-of-coordination curve shifts to the left (provided that costs of production are unchanged), which eventually shifts the overall cost curve to the upper

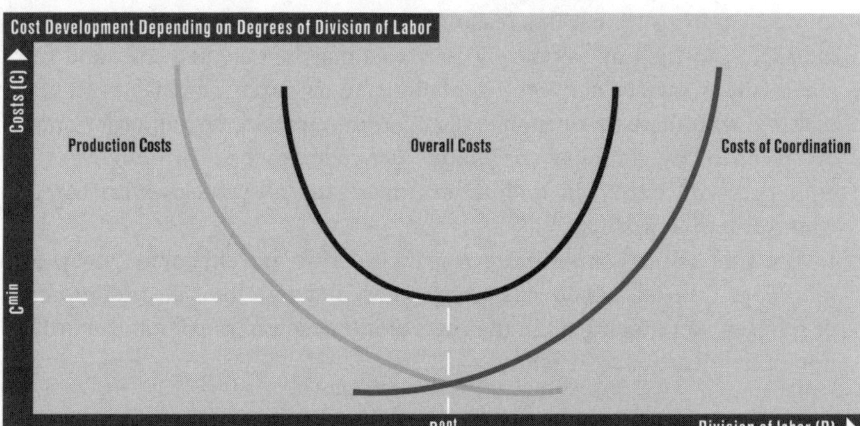

Fig. 4: Cost Development Depending on Degree of Division of Labor

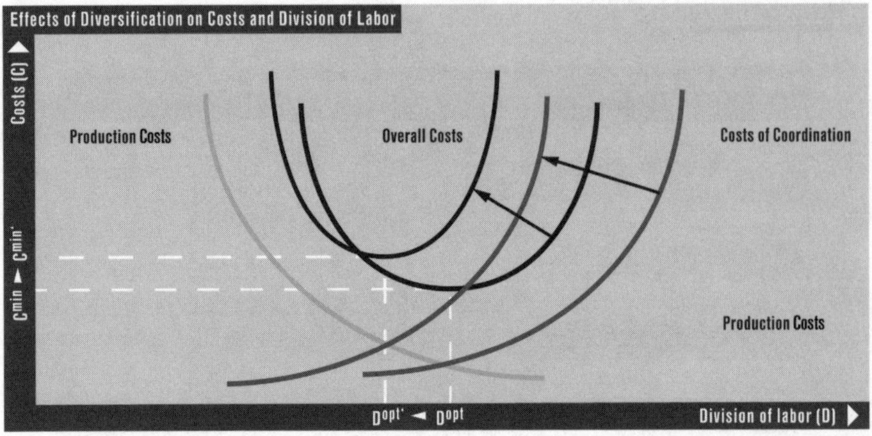

Fig. 5: Effects of Diversification on Costs and Division of Labor

left quarter. As presented in fig. 5, this equals an increase in overall costs and a shift of the optimal degree of the division of labor from D^{opt} to $D^{opt'}$.

Due to price competition, the increased costs resulting from customization are rarely acceptable for the average manufacturer, since they will hardly be compensated for by increased profit – even if we assume high prices can be achieved due to the customized offer. Two business solutions are possible:

The first solution involves a lowering of the costs of coordination, a process normally undertaken with the help of modern information and communication technology. This shifts the cost-of-coordination curve back to the right, suggesting that despite a high degree of customization, the curve indicating overall costs may be returned to its previous level or even to a lower one (see fig. 5). A few examples may demonstrate this development:

- Due to a considerable increase in its variety of models, Volkswagen expected the required stock for spare parts to increase from 280,000 articles in 2003 to 380,000 in 2008. For this reason, a new Original Parts Center (OPC) was installed, in which up to 70,000 deliveries may be executed daily and from where shops may be serviced twenty-four hours a day. This has been made possible with the help of mobile data communication, i.e., all order entries are recorded by stationary or mobile bar-code readers, all employees are equipped with handheld mobile terminals, and all vehicles with wireless communication systems.[41]
- In many e-business applications (media ordering, travel booking, etc.), configurators are used to help customers select and plan their desired products or services. At the same time, the costs of information transfer and coordination of delivery are reduced.

A second solution for reducing costs despite an increased degree of customization is to locate new divisions of labor, those in which lower coordination and

Turnovers and Major Customers of the Leading Companies in the Electronic Manufacturing Services

Company	Turnovers in 2002 (in USD approx.)	Major customers
Flextronics	14.4 bn.	Ericsson, Alcatel, Microsoft, Siemens
Solectron	12.6 bn.	Nortel, Cisco, Ericsson, Sun Microsystems
Sanmina	11,6 bn.	Dell, IBM, Hewlett-Packard, Nokia
Celestica	11.6 bn.	Motorola, Alcatel, Cisco, IBM
Jabil Circuit	3.9 bn.	Dell, Hewlett-Packard, Cisco

Fig. 6: Turnovers and Major Customers of the Leading Companies in the Electronics Manufacturing Services[43]
Source: Göbel and Molfenter: Arbeiten für die Riesen

production costs can be achieved. This is only possible if significant areas of value creation can be shifted to less-expensive suppliers without increasing – and perhaps even by decreasing – the costs of coordination. Again there are several business examples:

- The nominal level of vertical integration in the German automobile industry has fallen from 37 percent to below 25 percent in the last twenty years, and it is assumed that this development will continue.[42]
- Suppliers of electronic products such as mobiles, computers, handhelds, and game panels have also begun to outsource large sections of their production to companies in the so-called electronics manufacturing services. While the total worldwide turnover of these companies was 130 billion dollars in 2000, the forecast for 2008 is 164.4 billion dollars (see fig. 6).
- Large enterprises in the food industry have outsourced a significant part of their production.
- Hardly any firms in the consumer goods industry run their own transportation fleets. Transportation and accompanying logistics have been taken over by specialized service providers, often in the form of a full-service solution.
- Companies in the tobacco industry are currently considering outsourcing to suppliers of cigarette machines.
- Fashion designers like Escada have reduced their vertical level of integration to zero.

These and many other examples demonstrate that enterprises in the consumer goods industry are increasingly focusing on their core competencies, primarily in the areas of product development and marketing. Nearly all other operational areas are being outsourced to other organizations.

Effects on Value Creation in B2B Markets

Due to these trends in the consumer goods industry, we note two spillover effects on B2B business:

- value creation moves from the consumer to B2B markets;
- the question of the optimal division of labor in relation to the changed market conditions gains significantly in importance.

Since suppliers take over bigger shares of value creation, they are forced to discover new and more optimal forms of division of labor (upstream). Usually these forms result in the further outsourcing of value-creating activities. Suppliers have started referring to themselves as first-tier, second-tier, and third-tier suppliers; in logistics, there are lead logistics providers (LLP) relying on the services of third-party logistics (TPL) or fourth-party logistics (FPL).

A first observation that we can make regarding the new value chains is that value-creation processes, originally of an industrial nature, have turned into services. Often this means that B2B sellers start supplementing their products with services.[44] The industrial or functional services that are offered prior to, during, and after sales (pre-sales, at-sales, after-sales) are numerous.

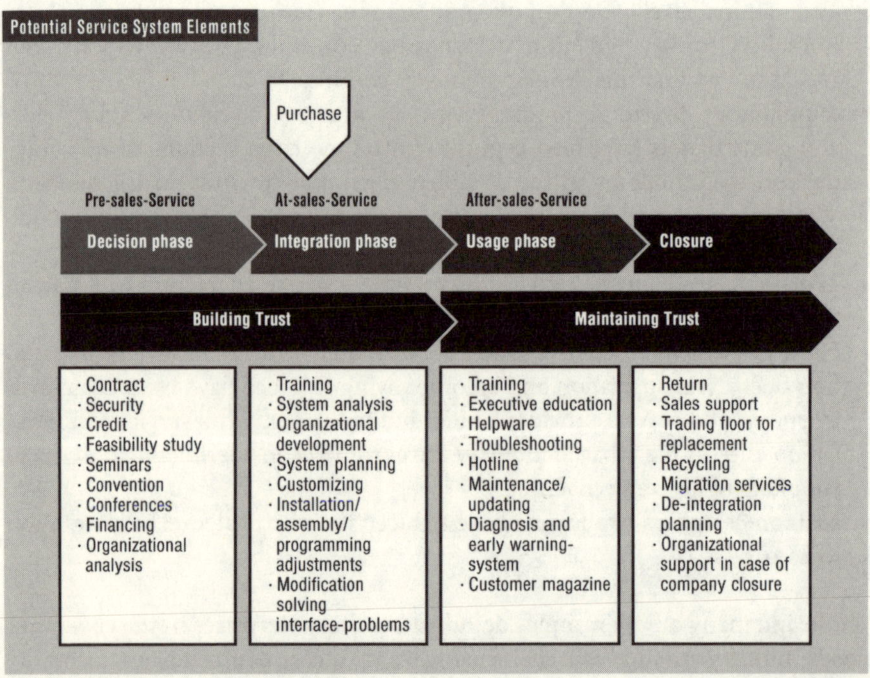

Fig. 7: Potential Service Systems Elements[45]
Source: Wimmer and Zerr: Service für Systeme

It is thus probably not surprising that different types of industrial service providers have emerged. Their spectrum ranges from simple *packagers*, who supply their customers with their own and related products, to *operators*, who, except in the area of marketing, assume vital value-adding tasks for their customers.

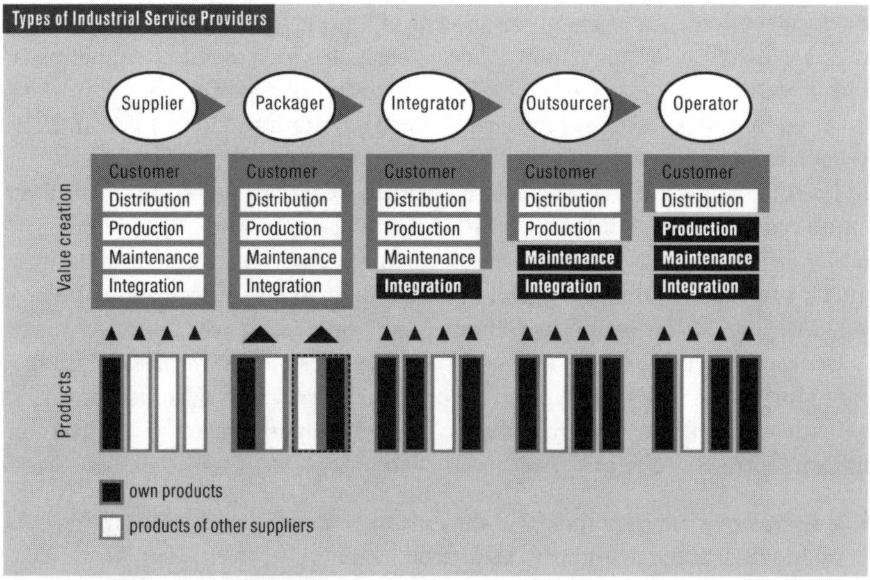

Fig. 8: Types of Industrial Service Providers[46]
Source: Monitor Group 2003

The term *operator* may already suggest that the service offerings of industrial manufacturers are becoming detached from their original product offerings. (The *operation* of a power plant has, for example, little to do with the production of its components and systems.)[47] In B2B business, such product-related services have been and will be offered not only by the product manufacturer, but also to a large extent as an individual performance by so-called *pure* service providers. Prominent examples are services for the telecommunication industries (in other words electronic data aggregation and processing), logistics, the leasing business, and the various consulting services, such as legal, tax-related, and business consulting, which are all chiefly performed for private and public institutions rather than for consumers.

The result of those trends is that B2B sellers have started handling their buyers' value-creating processes; which is to say, they have started providing services that may be sold individually or together with the product.

If we look at the three traditional employment sectors in Germany, namely agriculture, forestry, and fisheries; manufacturing and production; and the service industries, we will see that more than two-thirds of all employees are now working in the service sector.[48] Contrary to popular belief, the majority of these services are actually not performed for consumers. A close analysis of customer profiles in the different service industries shows that most services today are performed between companies[49] and are therefore part of the B2B market.

Statistically, the shift towards services has not yet been fully assessed. If, for example, we were to examine which tasks are performed by workers in the processing industries, we would observe that their work has equally shifted towards services. Today about 70 percent of workers in the processing industries are involved in either internal or external services, and a continuation of this trend is to be expected.[50] By the year 2010, the percentage of workers involved in producing goods in Germany will have dropped to 12.7 percent. The remaining workers will carry out internal or external service-related tasks.

The question that now arises is how such *modern* value chains, with their emphasis on services, differ from the traditional, product-oriented system. The major difference is as follows: whereas the previously predominant mass production needed no direct customer involvement, the services cannot be provided without such direct involvement.[51]

In order to produce services, the service buyer has to deliver *facts* – i.e., special information he offers the service provider or seller for a limited time and without which the service in question cannot be performed.[52] Such facts or information may refer to:

- the nature of the customer or of the customer's staff (for example in the case of an educational or training measure);
- objects, such as machines that need to be repaired or buildings that need to be cleaned;
- plants and animals to be cared for by an external service provider;
- legal rights that may be claimed by the service provider, for example a lawyer;
- nominal goods handed to a bank or company with the aim of gaining interest earnings; and, finally
- the information to be processed in the case of an advertising agency or a consulting firm.

A production involving the integration of the customer (see fig. 9) thus has the following characteristics:[53]

- the basis of the production is the seller's performance potential, which consists of resources (people, materials, etc.). With the help of these resources – i.e., prior to any order, simply based on speculation – the seller is ready to perform (e.g., to produce semi-finished and/or finished goods);
- external factors of individual customers are then integrated into the service-production process and the actual goods are produced;
- in the end, the customer receives a product that is a combination of various sub-elements established to varying degrees through the integration of the customer's own factors.

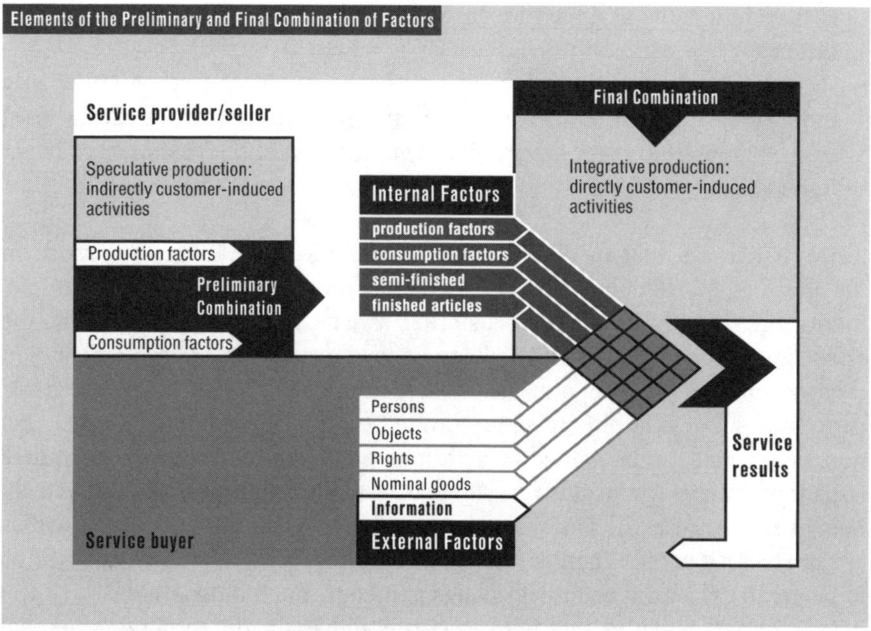

Fig. 9: Elements of the Preliminary and Final Combination of Factors[54]
Source: Kleinaltenkamp: Integrativität als Kern einer umfassenden Leistungslehre

The more customer-specific a seller's production or value-creation processes, the higher the seller's need to adjust his value chain. This degree of customer integration and adjustment is the factor affecting the B2B division of labor and the physical distribution of value-creating activities.

This has far-reaching consequences for the choices offered by suppliers active in B2B markets. Namely, depending on customers' decisions regarding their own performance guidelines, greater or lesser possibilities emerge for their (potential) suppliers to be used as such for performance areas that the customers themselves do not wish to handle. In this way, the possible needs of a customer-firm ultimately depend on its make or buy decisions. In order to understand this better and thus make it a basis for one's own market actions, we need to grasp the driving forces behind the customer's various outsourcing decisions.

Effects of Customer Integration on Division of Labor in B2B Markets

Whether a value-creating activity may be outsourced depends on two things:[55]

- first, on whether the activity is directly customer induced (and thus part of the final production) or rather serves to maintain readiness to perform (and

is therefore to be assigned to the speculative, preliminary combination of factors);
- second, on whether the value-creating contribution can be procured in a more or less standardized form from an upstream market or rather must first be tailored to meet the specific requirements of the customer or buyer in question.

Since all activities that are directly customer induced have a direct influence on the advantages sellers have to offer their customers, the quality of their implementation strongly determines customer satisfaction and, consequently, the sales figures. The process-related know-how necessary for implementation thus normally represents a valuable asset that a seller either does not gladly pass on at all or else only to a very limited extent. There is only limited economic advantage to be gained by a shift in process design; mostly it requires large investments (particularly so where information and communication technology are concerned). The result is that directly customer-induced activities are rarely outsourced. Their contribution to higher customer advantage seems to be greater than the potential savings gained through outsourcing.

In contrast, activities that help to create and maintain corporate potential have basically better outsourcing potential. They do not directly influence the offered service quality, but rather affect a company's overall efficiency. Here economic advantages through outsourcing are dominant.

Whether a certain value-creating activity is actually being carried out by another firm or not also depends on whether, and to what extent, a suitable problem solution exists. Although there are now outsourcing offers for virtually every need, it still makes a difference whether the activities in question *must* be or simply *may* be carried out in a customized or standardized way. If external partners are to render customized services, they will have limited possibilities to take advantage of economies of scale and synergy. If, however, they offer standardized services (i.e., more or less the same kind of services for a number of different customers), such advantages may be enjoyed and passed on as price advantages to the customer. Whenever value-creating activities can be outsourced in a form that is standardized, it can be expected that the positive influence of the outsourcing option on the efficiency of the outsourcing enterprise will be enhanced. This often applies to activities involving performance potential, since these are not directly influenced by the customer and may be better planned and thereby more easily standardized.

If both considerations are combined, the result is what is presented in fig. 10: B2B companies usually outsource value-creating activities that serve the preliminary combination of factors – i.e., creating and maintaining performance potential–and will most likely continue to do so. This is increasingly the case as these activities are increasingly standardized or can be standardized (square 1). A company is generally less likely to outsource activities undertaken in the final value-creation process, especially if there are no standardized solu-

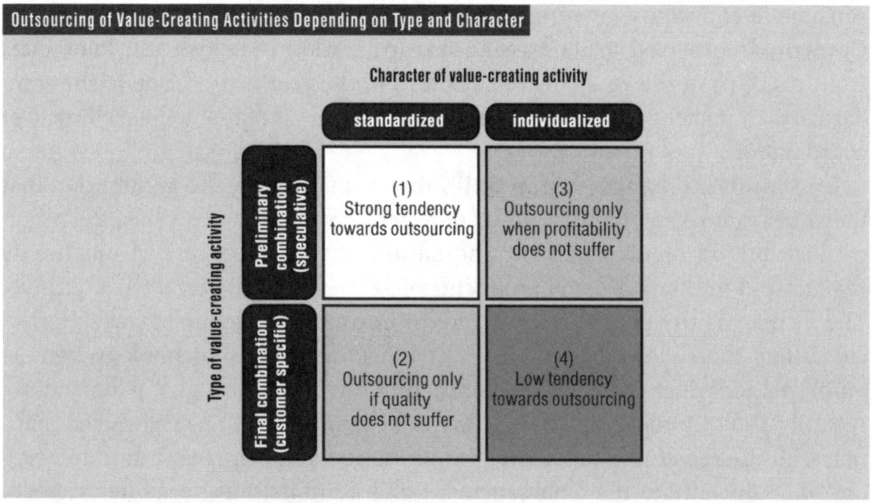

Fig. 10: Outsourcing of Value-Creating Activities Depending on Type and Character[56]
Source: Kleinaltenkamp et al: Outsourcing Kaufmännischer Dienstleistungen

tions available (square 4). However, the remaining two cases have to be differentiated. If an offer for external implementation promises a considerable reduction in costs, a seller can be motivated to outsource certain value-creating activities from the final value-creation process. (Call-center services would be an example.) In this case, however, assurance is needed that outsourcing will not impair the quality of the service offered, or that economic advantages are going to compensate or over-compensate these effects (square 3). Finally, highly specified activities in the preliminary combination of factors are outsourced when the resulting cost increases (due to customization) will not swallow the cost advantages (square 2). (This is often the case with specific logistical solutions.)

Effects of Division of Labor in B2B Markets on the Physical Distribution of Value-Creating Tasks

A general conclusion that can be derived from this discussion is that standardized goods are primarily outsourced to upstream market stages. The buying and selling of such goods tends to require low coordination efforts, so that seller and buyer need not necessarily be in each other's proximity. Finally, the physical distribution of these value-creating activities is determined by the cost of logistics. There is a tendency to physically outsource mainly traditional forms of industrial mass production to locations where their production costs are lowest–provided these targeted costs are not consumed by the resulting costs of logistics. As this trend allows logistics providers to achieve cost advantages, the radius within which the outsourcing of such tasks is economically sensible is constantly growing. It is thus not surprising that freight traffic in Germany is expected to increase in transport kilometers (per ton kilometer) from 169.9 m in the year 1990 to 485.0 m in the year 2015.[57] Due to the standardized character of the goods, there is basically no threat of increased costs of coordination.

In sum, those changes within B2B value creation allow the assumption that the shift from sectoral to functional or regional specialization[58] (i.e., the physical distribution of management and administrative tasks carried out in city regions and the standardized production activities in rural areas) will continue. The increasing use of information and communication technology will accelerate rather than slow this process. In most cases, this technology will be employed to reduce the costs of coordination incurred through customization. Actually, the technology itself will have to be configured in a customized manner, with the result that providers of information and communication technology will concentrate their management and administrative activities close to the location of their customers. If information and communication-technology solutions contain standardized components (i.e., for software programming) their production can be physically outsourced, just as in the case of traditional production processes. This tendency is reinforced by the fact that the results of such activities often take a digital form and can therefore be transported at low costs over great distances.

Conclusion

The increasingly individualized offers for goods and services on consumer goods markets have led to heightened complexities for the seller and resulted in higher costs of coordination. To avoid or reduce these costs, organizations in consumer and B2B markets are outsourcing value-creating tasks to upstream providers and sellers. This means that companies need to identify their core competencies (the ones that should not be outsourced).

At the same time, companies handling those outsourced activities have discovered and developed new business opportunities. Various options are feasible and can be observed; for example, previous pre-product suppliers who extend the range of their offerings, or new service suppliers who emerge, concentrating on the processes that have just been outsourced. Eventually, these companies as well will have to decide which of their new activities are going to become part of their core competencies and which not. They may in fact also come to reorganize their value chain and outsource certain activities or processes to another set of partners.

These developments require new and different forms of cooperation: to the extent that business processes are affected and altered, interdependencies between the players involved will be so as well. Insecurities regarding a new producer's ability and the quality of the outsourced product and production may arise. However, the decisions made here are of a long-term nature, and their results cannot easily be revoked, at least not the next day, not in a hurry. It is thus not surprising that the players look for reliable partners, for business relationships that can be sustained, rather than relying on contracts alone, which will never be able to cover all possible events. For this reason, there is a need for relationships in which partners know they can depend on and trust each other, however difficult a given situation may be.

This results, in turn, in consequences for the tasks of the sales division and the development of sales activities. Increasingly, the individual sales representatives have to develop from sellers to business relations managers. But since the above-described decisions on the customer side are usually not simple purchasing decisions but are mainly made by other, mostly higher-placed persons, customer care needs to adapt accordingly. The buying center becomes a relationship center, with the result that on the supplier side adequate structures likewise have to be built up and managed. What emerge here are changes not only in the development of key-account management system, but also in the choice and development of the personnel performing in these areas.

In addition, multi-stage marketing activities take on ever-greater relevance. For if the decisions of the direct customers depend on their own orientation towards the value-creation structures of the customers' customers, one significant factor involved in the success of sales activities emerges: the decisions of this group need to be influenced in such a way that there is (must be) a demand for the direct customers' own products and services.

Such multi-stage marketing involves all sales-related measures aimed at the market stages (customers of the customer) that follow one or several direct customers. A multi-stage sales strategy thus always targets at least one subsequent sales stage. It can, however, also comprise several stages. The target groups of multi-stage marketing can be made up of all processing and trade stages that follow the direct customers, down to the stage of the final user. It should be additionally taken into consideration that decisions to purchase are not only made by the organizations or individuals within the direct processing chain themselves. In the different stages, the influence of other individuals or organizations within the environment of the market stage can take effect.

A comprehensive, multi-stage marketing plan is therefore characterized by a network of all different kinds of causal relationships. This is presented as an overview in fig. 11.

In a multi-stage marketing plan, an individual marketing mix is made for each stage or target group a seller is eyeing that takes the customers' interests into account. In this way, all relevant sales stages are tied into a single compact and comprehensive marketing strategy. Finally, the method of approach is aimed at transforming adversary market relations between the stages into a more cooperative form. All market stages should be in the same boat, so to speak, and each stage should make its own contribution to a vertically coordinated strategy.

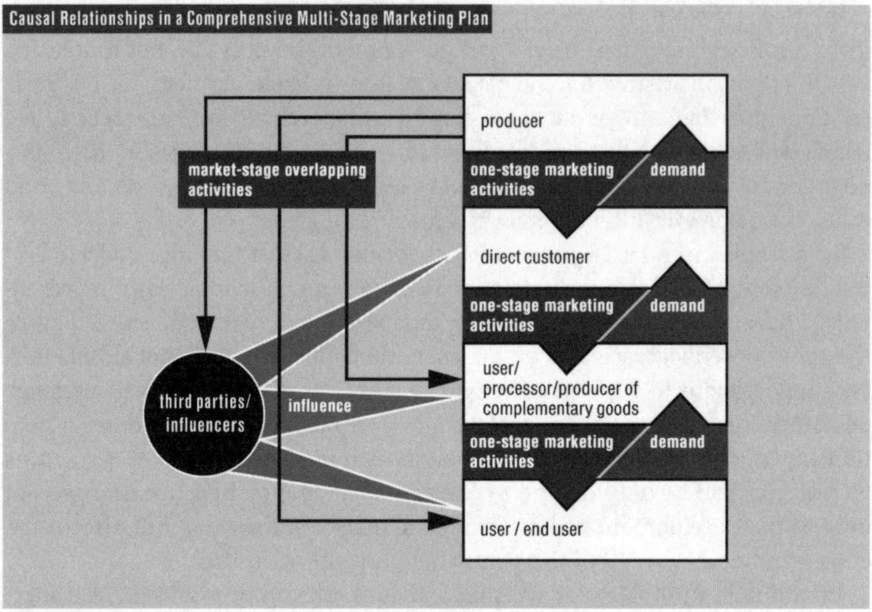

Fig. 11: Causal Relationships in a Comprehensive Multi-Stage Marketing Plan

In order to achieve the desired effects of a multi-stage marketing strategy, it must however be clear to customers in downstream market stages that the upstream goods and services are of significance for their own competitiveness and that they can influence their own profit. If there is no awareness of this connection among the customers, or if it is not made clear to them, then multi-stage marketing loses its purpose and is doomed to fail.

4
Cost Decisions and Pricing Decisions in Times of Value-Based Management

Mario Rese

Innovations, Competitive Advantages, and Value-Based Management

At present, bringing new technologies, goods, and services to the market in an ever-faster manner is a necessary albeit by no means sufficient condition for a company's long-term survival. What is also required is that the innovations furnish sufficiently positive earnings. In this regard two questions come to mind: First, in the present period of globalization, how can sustainable advantages in competition be realized? And second, how can we efficiently define both potential and realized earnings, and to that effect measure them before as well as after?

This chapter is concerned above all with the second question – that is, with the measurement of the economic results of innovative goods and services. At first view the answer here may seem very clear and simple. What determine the added value of an innovation are both the incoming payments it has prompted and the accompanying out-payments over its full life cycle. In accordance with their temporal accrual, these in-payments and out-payments are discounted at a rate fixed at the point of purchasing decision, with use of a WACC computation (= weighted average cost of capital: costs of the capital tied up in the company). The value of a new product, service, or technology is assessed on the basis of an NPV (net present value) computation. Positive NPV signals a rise in the company's value; this also serves as a signal to actually undertake the investments connected with the innovation. In contrast, negative NPV signals a loss of company value, with the entire innovation-project not being realized as a consequence. This follows the proviso that every investment had to bring in at least the company's capital costs.

Alongside this well known investment perspective, there is, however, a second approach to evaluating success, one placing earnings achieved within a specific period in the foreground. Within this period-centered framework, it is also expected at present that all of a company's activities (in sum) generate EBIT (earnings before interest and taxes) that in relation to the tied-up capital achieve a ROCE (return on capital employed) that turns out to be higher than the WACC. Then and only then do we have positive EVA (economic value added). From the period-oriented perspective, present-day so-called value-based management emerges from the EVA, ROCE, and WACC triad.[59]

Bringing Technology to Market. Olaf Plötner and Robert E. Spekman

ISBN 978-3-527-50270-7

When combined, these two approaches to company success furnish a very clear picture of the monetary consequences of innovative technologies, products, and services. The investment perspective focuses on the isolated project through its entire life cycle; for its part, consideration of the period involves an assessment – mainly in the form of an internal computation of costs and revenue – of the project's contribution to the company's total change in value for the business period.

If we pose the question of when which computation of success is used, then it becomes apparent that consideration of the project has significance above all before the decision for or against the entire project. In the rule, the focus here is on "go or no go" regarding a new product, service to be offered, or technology. On the other hand, the cost or revenue-oriented perspective is deployed for operative management following a general "go" decision for the project. Questions are answered regarding whether a certain customer or contract will pay, what the pain-threshold is for still accepting a contract, and so forth. The following observations will be concerned above all with this second, operative perspective: what instruments are available for the period-oriented management and evaluation of the success of technology- and goods and services-related innovations, and how well do these instruments function?

At this point the critical reader may observe that books on management accounting are full of recommendations on this theme, thus implicitly raising the question of this chapter's added value.[60] What is new in this context is related to the fresh challenges being posed by globalized markets. We begin with the assumption that in the future above all firms from the established industrial nations – Japan, the United States and Canada, Western Europe – will have to change their strategies of production and innovation. Standardized goods and services alone are increasingly ill suited for generating sustainability, no matter how superior their efficiency at the time they were developed. The speed of imitation has accelerated and will continue to do so, with the result that precisely established providers that are strong in research will receive their investments in ever more amortized form. The time spans for skimming off innovation-based profits are becoming shorter, and in the global context protective mechanisms such as patents and utility models are losing importance.[61]

Concepts are needed for avoiding or at least slowing down the competition's imitative efforts. One term that will have a magical aura in the future is "industrial product-service systems" (IPS^2). The integrated development of service-product mixes tailored to individual customers could generate entirely new barriers to imitation, allowing a company more long-term advantages in developmental work.

But until now no adequate value-based management instruments have emerged for such IPS^2s. Above all on the level of operative decisions, the known instruments (direct costing, activity-based costing, and so forth), applied in isolated form, end up grasping at air.

This describes the present chapter's framework. The following section will offer a more detailed look at the idea of the IPS²s and the arising challenges for firms developing such systems. On this basis, the third section will focus on the question of pricing these IPS²s.The goal must be to move the price as close as possible to the client's actual willingness to pay – without, however, ending up losing the client. The fourth section centers around the question of price floors: What minimum fees have to be earned in an IPS² in order to not end up worse off then without the contract? Or posed differently: where does the price floor lie in an IPS²? In this respect, the discussion will not only treat possible measuring methods for price floor and price ceiling but also the management of both these decisive benchmarks. Creating the greatest possible distance between floor and ceiling raises the probability of actually contributing value to the company with IPS²s.

Sustainable Competitive Advantages in a Dynamic Business Environment: IPS² as a Source of Success

One unshakeable maxim of business practice is: competition is dynamic. Those incapable of adapting to constantly changing competitive conditions will not survive. Over recent years, above all firms from the Western industrial countries have had to painfully experience this truth. Until the 1980s, for national and international success it was mainly enough to transform superior technologies into product-solutions. But the increasing international integration of economic streams has led, among other things, to technical disadvantages on the part of firms from so-called emerging markets themselves being overcome at an ever-faster pace. For the established providers, this means – when taken together with their massive disadvantage in wage costs – that many of the goods and services being delivered are hardly still competitive. Encouraged by their initial successes, increasing numbers of firms from China, India, and South America have begun to discover increasing fields of business and to assert themselves against Western competition in many branches of their home markets. And more: they have been having increasing success in marketing their own products in the Western industrial countries. For where in China average hourly wages are approximately 0.9 dollars, and even in the Czech Republic the figure is only 3.4 dollars, in the Western industrial countries it is running between approximately 20 dollars (USA 22.1; Japan 20.5; France 21.5, and the United Kingdom 19.5) and 30.5 dollars (Germany) – and this with up to 25 percent less work time per week.

Against this backdrop, in order that firms in the Western industrial countries remain competitive, disadvantages in labor costs need to be compensated for through higher work productivity. It is above all in respect to that index that the Western countries still maintain a lead over non-Western competitors such as

Vietnam and China. Where, for instance, the GDP for China is only 10,000 dollars per employee, the leading industrial nations show figures between 59,000 dollars (Germany) and 89,000 dollars (USA). But it is also the case that if we do not manage to carry on developing business models offering customers superior goods and services, sustaining the advantage will be difficult.

In many branches, one initial reaction to this threat has been a form of service engineering: increasing augmentation of goods-related marketing activities with preliminary, collateral, or downstream service activities. But because of the intensity of the wage costs involved in many product-accompanying services, together with the often only tepid willingness to pay for these additional services, at present this expedient tends to lead to a dead end. In addition, inventive entrepreneurs are succeeding in offering the services without the base product, thus frittering away a possible competitive advantage through service supplementation.

In this regard, one highly promising solution could be combining goods and services into an IPS^2 in such a way that separating these components is inherently impossible. This requires a combination of goods and services no longer viewed as based strictly on provision of either one or the other – on the services being conceived around the goods or goods coming with the service provision. IPS^2s are stamped by an integrated and mutually determining process of planning, developing, provision, and usage of goods and services. In this way the customer is offered new and to large extent initially unknown possibilities for solving problems. A partial substitution for the goods-based and services-based components of the IPS^2 is here made possible; the provider can introduce it as a formative component to satisfy the individual needs of clients.

An IPS^2 can be composed of various components: either additional hybrids – that is, modules combining goods and services–or pure service-provision modules, or pure goods-provision modules (see fig. 1). The factors determining the correct mix of goods and services are the state of knowledge and the customer's capabilities. What counts is as follows: only offer customers the goods and service provision modules that they themselves do not possess or control. As an example, we can imagine a manufacturer of specialized machines who offers his customers a range of contract variants according to their know-how regarding deployment, operation, and maintenance. These variants extend from performance contracting in the sense of full-service contracting including fabrication of the necessary parts to the pure delivery of the machine with a simple spare-parts service where the responsibility for both fabrication and functions lies with the customer. An offer is thus extended according to the customer's capability profile, the absent goods and services components being delivered according to the lock-and-key principle.

In order that the provider always offers the best goods-services combination for individual customers, appropriate knowledge of their processes and process capabilities is called for. It is clear that in the presence of a heterogeneous knowledge basis on their part, we need to speak of individualized industrial

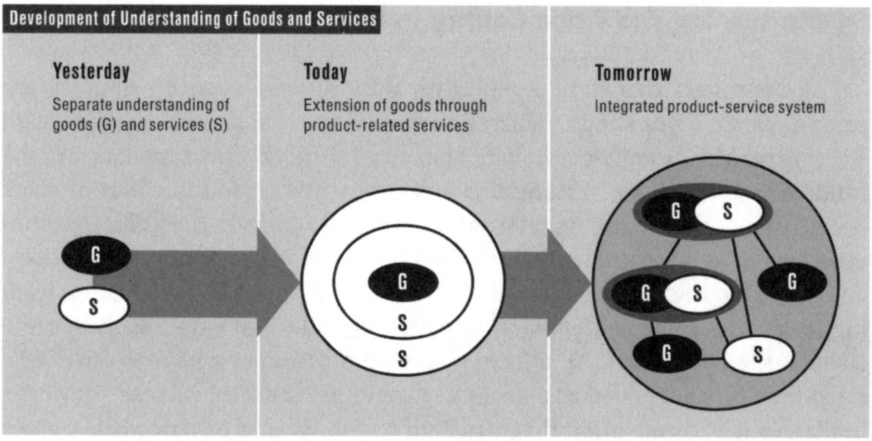

Fig. 1: Development of Understanding of Goods and Services

product-service systems. Configuring these properly presumes a deep understanding of customer processes and the customer's knowledge base. And the desired sustainability effect is located precisely at this point. The provider's competence is no longer located solely in an ability to generate attractive goods and services. It is now also – and above all – located in the ability to recognize customer problems and furnish the best individualized answers in a framework of tailored IPS^2s.

This creates advantages for the IPS^2 provider. In the first place, the individualized IPS^2 leads to a diminishing of the threat posed by production-cost advantages on the part of competitors from the emerging markets. In the second place, the integration of customer and provider processes leads to a diminishing of the provider's replaceability. Process integration, understood as a linkage of the customer's value chain with that of the provider, has the effect of barring entry from the outside into the customer-provider relationship. And finally, the capacity to analyze the customer process and identify the sort of goods-services modules needed is a competence not easily reproduced in the same way by other, emerging market providers. Hence particularly these knowledge-based capacities have been ascribed with both high sustainability and relatively more difficult imitability.[62]

All told, we see the future of firms from the triad countries as lying mainly in such IPS^2 solutions, with their integration of goods and services. Individual industrial firms have already recognizably taken this direction. The question here emerges of whether our well-known management instruments are suited to accompany and support this process, above all in the realm of cost and revenue analysis. This will now be clarified step by step, first by focusing on the question of identifying willingness to pay.

Determining the Price Ceiling in IPS²s

From a purely monetary perspective, the customer's price ceiling is the price at which he or she is prepared to either switch to another supplier with a comparable solution, or simply not go ahead with the investment, or use the funds elsewhere. If we assume that customers reveal the same sort of value orientation in their thinking referred to above, two critical price-ceiling parameters can be identified. The first of these is located where customers precisely do *not* earn at least their own capital costs through the IPS² – a situation leading to an unacceptable depletion of company value, with the result that the customer fails to invest. An offer providing the customer with a positive NPV must thus be understood as a necessary condition for sales success. However, in the event of competition this is not sufficient. What is also needed – hence our second parameter – is a comparison with the NPV of the most dangerous competitor. The NPV of your IPS²s has to be at least marginally higher than that of the competitor. The customer will purchase from you in that case – and only in that case. What this means is that the price we determine must involve a small, carefully calibrated undercutting of the competitor's NPV while at the same time generating positive customer-NPV. Together, these two conditions constitute the benchmarks around which our pricing policies need to be oriented.

It has become clear that the upper price limit depends on the customer's expectations regarding costs and benefits. To understand these expectations the seller needs to know what the customer intends to do with the IPS². We can here distinguish two basic situations: (1) The IPS² leads both to revenue (hence to payments) and to costs. (2) The IPS² only influences the company's cost situation. The processes are altered with corresponding (hopefully positive) consequences for the customer's costs.[63]

In the first case, to judge an investment's benefits it is necessary to estimate both the revenue side and costs side. This furnishes the specific payment series of the IPS². The NPV of an investment is defined from a customer's perspective as

$$NPV_0 = -I_0 + \sum_{t=1}^{n} (R_t - E_t) \cdot (1 + wacc)^{-t}$$

with: NPV_0 Net Present Value in t=0

I_0 Investment in t=0

R_t Revenue (inpayments) in t

E_t Expenses (outpayments) in t

wacc Weighted average costs of capital of the customer.

This means that the NPV_0 at time 0 (time of contract) is equal to the discounted value of the net income stream (income R minus expenses E) from the IPS²'s use over the period 1 to t, less the initial investment/payment I_0, the purchase price.

For the sake of simplicity, we will initially assume that the supplier has no competitors and the customer decides whether or not to invest with the supplier. The price ceiling is where the NPV is zero, that is, the discounted net income stream equals the purchase price:

$$0 = -I_0 + \sum_{t=1}^{n} (R_t - E_t) \cdot (1 + wacc)^{-t}$$

If competitors exist, the theoretical upper price limit is determined by the strongest competitor. Let P^S be the price of the focal supplier and P^C be the price of the competitor. R^S and E^S are the revenue and costs/expenses per period for the focal supplier of the IPS², and R^C and E^C the revenue and costs per period for the competitor's solution. The differential advantage of the supplier's IPS² solution compared with the competitor's is defined as:

$$NPV_0^S - NPV_0^C = -(P^S - P^C) + \sum_{t=1}^{n} [(R_t^S - R_t^C) - (E_t^S - E_t^C)] (1 + wacc)^{-t}$$

Setting the NPV difference to zero and solving for P^S gives the price at which the customer considers both suppliers equal:

$$0 = -P^0 + P^C + \sum_{t=1}^{n} [(R_t^S - E_t^S) - (R_t^C - E_t^C)] (1 + wacc)^{-t}$$

$$P_{max}^s = P^C + \sum_{t=1}^{n} (CF_t^S - CF_t^C) \cdot (1 + wacc)^{-t}$$

with: $CF_t = (R_t - E_t)$

To express this in words, the focal supplier's price can be greater than the competitor's to the extent that the discounted net income stream for its IPS² solution is greater.

Often, income flows cannot be determined. This is the above-described case (2). In such a case, it is possible to compare the costs incurred over the life of the IPS² with the costs incurred by the competitor's solution:

$$NPV_0^C - NPV_0^S = -(P_0^C - P_0^S) + \sum_{t=1}^{n} [(E_t^C - E_t^S) \cdot (1 + wacc)^{-t}$$

$$P_{max}^s = P_0^C + \sum_{t=1}^{n} (E_t^C - E_t^S) \cdot (1 + wacc)^{-t}$$

This means that the focal supplier's price can be greater than the competitor's, depending on the discounted value of the cost savings. It becomes apparent that this is nothing more than the total-cost-of-ownership approach with a dynamic component.[64]

As clear and logical as this argumentation may seem, it invites two questions: (1) Do customers really think like this? (2) Can a provider know the needed input data well enough to actually be in a position to use these computations?

On (1): In actuality, customers will not generally make such life-cycle advantage computations. For many, prices continue to be the main consideration.

Others use signals such as brand names, trusting that they are then on the safe side in their purchasing decisions. And yet others simply trust their feelings.

It is nonetheless also the case that in a competitive context that is growing more difficult, entrepreneurs have to increasingly rely on economic criteria in making decisions if they do not wish to find themselves eliminated in the long term. And this is the point where the proximity to reality of such a price ceiling determination comes in. If we assume that the input data can indeed be estimated in a realistic way, the process reveals itself as an instrument for both persuasion and selling that presents customers with the direct economic effects of their potential purchasing decisions. Even those who have never employed such computations are very likely to find their results of interest. And this is all the more the case to the extent their markets are facing strong competitive pressure.[65] At the same time, this tool can be used in sales conversations in order to interactively develop possible future scenarios with the customer. In a situation where future sales are not specified as a fixed value, but rather where several background scenarios, each with its subjective probability of occurrence, are considered instead, then a highly individual picture of the customer's subjectively perceived business situation can be formed, with a corresponding assessment of the investment that needs to be effected. There is certainly a possibility of this contributing to the emergence of a great deal of confidence in the sales manager. In addition, in the sales process arguments coached in quantitative, economic terms are often much more forceful and difficult to refute than purely qualitative arguments. This tool thus offers the sales manager a chance to develop both the power of numbers in dollars or euros and an entirely different kind of mutual trust.

On (2): In reality, significant information problems arise, including competitors' prices, operating costs, and income streams, with the result that no one is able to estimate the real future NPVs. But this constitutes anything but a methodological knock-out argument. For what is at stake is not the possibility of knowing everything correctly, the interesting questions rather being: at any given point, who is best informed about the relevant information, and how large is the gap in knowledge of the provider making the computations?

If the computation can and should be an instrument of persuasion, being plausible and credible is sufficient. Complete accuracy is not a demand that can actually be fulfilled. To this extent, what needs to be considered is who really knows something about the separate components and how good this knowledge is in comparison to one benchmark: the person who is best informed in each particular case in the market. From the view of the provider, the components are as follows: (1) one's own future payment structure E_t^S. It should be assumed here that even with knowledge by customers of the conditions for usage, the provider can best estimate what the payment series will look like. (2) Customers should be best informed regarding the revenue R_t^S attainable with the IPS^2. But at the same time sales managers of providers should themselves be well informed about the conditions to be expected in customers' markets.

Ideally, good sales managers will know the risks and opportunities facing customers in their markets as well as the customers themselves. (3) Competitors are themselves naturally best informed regarding their cash flow. But it is also the case that using available observation tools, a good sales team has solid knowledge of the assets and deficits of the competition's solutions, extending to maintenance cycles and so forth. (4) Finally an estimate needs to be made of the competitor's price. Here as well the advantage lies with the competition. But it is also the case that a sales team is familiar not only with the competitor's usual price setting but also with factors such as the degree of aggressiveness regarding discounts, as manifest in a particular situation. If the competitor is doing well, with filled order books, expanding markets, and moderation on the part of its own competition, less aggressive pricing can be expected. On the other hand, if the competitive context is difficult, marked by disused capacities and so forth, pricing aggressiveness will be markedly stronger. In order to deliver a halfway realistic prognosis concerning the competition's behavior, the sales team needs to know all this, at least on a basic level – even without use of an NPV computation.

All told there are very good arguments for using this method. The fascinating thing here is that it has direct bearing on the customer's most salient problem, at least that of the value-oriented customer: what changes in company value can I expect through an IPS^2? This tool's attractiveness is further underscored by citing two additional applications: its usage for provider-internal identification of price margins; and its external usage as a sales instrument directly with the customer.

We need to now recall the specific goal being addressed in these pages: following a basic decision for a new product, service, or technology – in our case tied to the IPS^2s – to achieve the highest possible degree of raised value on the provider's side or at least keep value depletion within the narrowest possible bounds. Speaking very generally, we here have three variables: (1) the price for an IPS^2 in a specific sales situation, as discussed in this section; (2) the costs of the IPS^2s; and (3) the number of sold or produced IPS^2s per customer – or customer group – and period.

The following section is concerned above all with the question of IPS^2 costs relevant to purchasing decisions. These are important because the particular value contribution emerges (after subtraction of interest and taxes) from a comparison with the price. And multiplication by that quantity determines the absolute value contribution, measured with the key performance indicator EVA.

Determination of the Decision-Relevant Costs of IPS²s

Preliminary Considerations

From a management perspective, determining the costs of goods and services is one of the most basic processes; it is also a process often cast into doubt. Precisely between controlling and sales, very different views frequently emerge regarding the true extent of costs, or urgency or necessity of the coverage of various cost components. Present to greater or lesser degree in every firm, this conflict has its source in the controller viewing company success as a central goal – one all the more assured the fewer pricing concessions sales offers to customers. Defined as a border that should not or must not be undercut, costs are here often very welcome as motivational help offered the sales manager in not acting too gently vis-à-vis the customer. But in fact, this justified or sometimes artificially generated cost pressure does not make things especially easy for the sales manager. And if the customer remains firm regarding a desired price reduction, and the sales manager's incentive comes from either turnover or earnings/contribution margins, it is more than understandable to at least make a try at calling into question the cost figure determined by controlling – whatever that figure may be.

Beyond this conflict, everyone agrees that at least in the realm of mass and serial fabrication the full costs, in other words the individual costs plus the apportioned overhead, are hardly adequate for judging an individual business. At least in the short term, from a decision-oriented perspective only the costs that are still influenced by the decision for or against the contract should be considered as a price floor. For this reason, they are called relevant costs. But it is equally clear that in the long term, all costs have to be at least covered by all revenue to achieve positive company results, and beyond that a value contribution for the company. To this extent, the long-term price floor is present in ideas on value orientation (see section 1 of this chapter).

The theme being addressed here is part of a discussion of the "correct" short-term price floor. Many publications have contributed to this discussion – above all in relation to the factors influencing the height of the floor.[66] In this chapter the focus will above all be on the question of how the short-term price floor for IPS²s can be identified. The substantive problem at work here lies in the fact that we are speaking of always-varied mixes of products and services, each mix demonstrating a very different cost structure. Together with the idea that through IPS²s providers and customers are frequently linked in a business relationship of a more long-term nature, this produces a no-longer trivial problem of determining the economically sensible price floor.

Proceeding in several steps, we will first consider the problem of price floors in relation to products. We will then turn to services. Finally we will discuss possibilities of integration, in order to do justice to the special characters of

IPS²s. Here the effect of customer relations on the height of the price floor will be examined and integrated into the decision framework. The focus will consistently be on the sales process.

Direct Costing for Products

When it comes to determining the short-term price floor for products, the specialized literature is of one voice: as indicated, from the perspective of the single transaction the only costs that must be covered are relevant costs, those still capable of being influenced by the contract decision. These are either variable costs or individual costs, depending on the sort of business. In batch and bulk business, it is best to work with variable costs; in project-based business possessing a high degree of individualization and low batch sizes, working with individual costs is preferable. The basic requirement is that a contract has to at least cover the variable or individual costs. Then and only then is it assured that revenue exceeding these relevant costs can also cover a portion of the in any case accruing fixed or overhead costs. On the other hand, an undercutting of the relevant costs would mean that in realizing the contract the provider would be worse off than without realizing it. In other words: if the revenue undercuts the relevant costs, it is economically damaging to accept the contract.

This basic idea has been frequently discussed and relativized in different directions.[67] But this does not shake the idea's essential correctness. Nevertheless, the question is present of whether in the case of IPS²s the approach is still tenable or perhaps needs expansion. It is a fact that over the past twenty years the portion of overhead and fixed costs in the total costs of a producing industrial enterprise has increasingly risen. At present, for industry we assume an average overhead share amounting to two-thirds of the total costs – a pronouncedly larger share of this being fixed. This means, inversely, that the portion of relevant costs in the total costs is increasingly smaller. A potential contribution margin calculated as the difference between the revenue and the relevant costs is ever larger or, more precisely, increasingly equal to the revenue. In an extreme case such as the telecommunications industry, we can hardly any longer identify costs that are contract-related or customer-related, with the result that the relevant costs and with them the short-term price floor tend to move towards zero. Put more simply this means nothing else than that at least from a costing viewpoint more or less any price would be acceptable. This is not desirable for various reasons:

(1) At the end of the period all costs including those that are not relevant have to be at least covered if the company is not to incur any losses. For this very reason covering the relevant costs alone is insufficient.
(2) A completely free pricing policy, following the dictum "the main thing is that the firm sells over the price floor," destroys the price level in the heads

of the customers, with devastating consequences for the long-term profitability of a field of business.

(3) The decision-supporting hence behavior-steering effects of the cost information completely vanish. Marketing lacks argumentative potential for defending its own pricing concept in face of the customer. It loses the feeling of needing to cover the not directly relevant but nevertheless accruing overhead or fixed costs.

These brief points already suggest some consequences for management: the larger the portion of individual costs or variable costs in the total costs, the more significant relevant costs become for estimating the price floor. Inversely: the smaller the portion of individual or variable costs in the total costs, the less helpful the knowledge of the actually relevant costs for making decisions in markets regarding price assessment. In addition, revenue interrelations such as those in business relations lead – speaking very generally – to price floors determined by cost-calculations alone not necessarily being correct. Their undercutting in order to save the customer relationship and hence future revenue from the customer is indeed advisable. In sum, at least in management practice the price floor that is correct in terms of cost calculation does not have to correspond with that which should actually be determined.

In relation to the IPS^2s, one more consideration needs to be mentioned. Precisely for the services sector, it is the case that the accruing costs reveal no genuine direct cost character in relation to a contract or a customer. And on account of the high wage quota and usually rigid wage structures, they can only be described as variable to a limited degree. For consideration of price floors, such predisposed costs, no longer capable of being influenced in the short term, are irrelevant. This suggests that precisely in the case of IPS^2s, depending on the goods-services mixture and the variable or direct costs accruing here, at least in the framework of a contract or customer only very small relevant costs exist. For the reasons suggested, a short-term price floor determined in this way would offer no help in deciding, hence would be practically useless.[68]

Against the backdrop of common practice, it would also be inadequate. Particularly in the present-day service industry, it is possible to react very quickly with company capacity. When a specific service does not pay, in other words when the revenue lies under the total costs (capital costs considered), the response is sometimes a rigorous restructuring and downsizing, the service in question no longer being offered. Because of this ability to react in relatively short-term fashion, at least in an extended sense part of the costs of a service should be considered relevant. However, in order to be able to take account of these service costs in a determination of the price floor, a mode of allocation needs to be found allowing identification of service costs per contract or customer.

In many industries, this is taken care of entirely through entry of hours. This makes sense when there is a clear connection between the employees working on behalf of a contract or project and just these allocation objects.

In more long-term projects with clear assignment of a team of employees this is frequently possible. But this is not the case with marketable service modules, in which other costs as well accrue – driving costs for company cars, tool wear and tear, and so forth. Here activity-based costing is very often recommended. This allows a charging of the service costs corresponding to the frequency of certain realized activities – which in turn makes a gauge available for allocating activities and processes to contracts and customers.

The basic features of this form of costing will be reviewed in the next section, in order to develop a decision-relevant price floor for the service sector of our IPS^2s as well.

Activity-based Costing for Services

Activity-based costing allocates activities or processes to overhead costs in order to then charge these activities or processes to purchase orders or customers for their usage: An *activity* is a repeated, identical sequence of actions in firms that generates costs and to which these costs can be allocated in the causally most suitable manner (as individual costs of the activity or else allotted as pro rata overhead activity costs). Very frequently, several activities are brought together into processes. One example of the result is a stock-withdrawal process consisting of the activities "fork-lift trip to the warehouse; removal of shelves; return of fork lift; documentation of removal process." When it is sensible several processes can be brought together into so-called primary processes.

A condition for carrying out activity-based costing is that repetitive activities are actually present: that the activities – and thus the process – are carried out again and again in a nearly identical manner. In the framework of the calculation, a contract or a customer is then charged for the costs of the various processes to the extent these have actually been accepted for usage.

Activity-based costing consists of three steps: (1) Ascertaining the process costs; (2) determining a process-cost charge; and (3) charging the process costs to the contracts or customers.

In the first step, the costs directly attributable to an activity – for instance the salary of a service employee – are indeed allocated to that activity. If such a cost item covers several activities, distribution among the activities needs to be undertaken, in the form of a pro rata cost-overhead allotment. What results from this procedure is savings on the part of the classical cost centers, because from now on a part of the overhead is charged to the activities. In this way, with use of activity-based costing the level of the surcharges sinks. In the second step of determining the process costs, the costs per process need to be identified. This requires first asking which activities can be brought together with which processes. Then the overhead costs for the period per activity or process are added up. This produces the process costs per process and period. Dividing by the number of processes requested in the period under consideration produces the rate of process costs. In the last step the process costs are charged to the purchase orders or customers. For this it is only necessary to multiply the number of processes that have been accepted contractually or by a customer by the particular rate of process costs. The result is that an order or customer is charged with exactly the costs that have been activated through acceptance for usage of the particular process.

In order that the process costs can be understood as nearly relevant costs, they have to be available to a certain degree. This means that if, with consideration of the processes produced by a service, it turns out that the service does not pay, it has to be possible to reduce or eliminate the costs tied to the process as quickly as possible. Then and only then can the process costs be viewed as a rational price floor for services.

In the framework of the IPS^2s being considered here, this means that the individual service activities need to be available in as mutually independent a form as possible. Then the process costs can be viewed with a good conscience as a short-term price floor for services.

For an IPS^2 the total price floor is composed of the individual costs of the hardware component plus the prevailing process costs (available on a very short-term basis) of the integrated services. For determining the price floor of IPS^2s, we thus use an integrated direct costing/activity-based costing approach. This allows the relevant costs of both the goods and services components to be considered. This is no perfect solution. But it is distinctly better than the isolated use of either the direct costing or activity-based costing approaches.

In any event, there is one more systematic problem in determining the price floor of IPS^2s – a problem not yet adequately treated in our discussion. In the case of existing revenue associations – for example those resulting from established business relations – in view of a direct transaction it can be highly recommendable to undercut the actual price floor (the direct costs of the contract plus process costs of the service). The explanation for this is that the future of the entire customer relationship can very much depend on realizing the contract. If future contracts (hence the customer) are presently viewed as raising company value, then the price floor, understood as the lowest acceptable price, can lie distinctly below the relevant costs. Economically the level of the future customer net present value determines the maximum acceptable undercutting of the relevant costs in the current contract to be decided.

The following section will consider this final problem in determining price floors.

DC-ABC for Integrated Product-Service Bundles in View of Revenue Interrelations

In a company, once the hard, quantifiable arguments are exhausted, strategy comes into play. The talk then turns to strategic projects and strategic customers. There are demands for references and exhortations for strategic decisions.

It is the case that a range of valences regarding customers exists. It is also the case that the associative relationships between individual businesses need to be considered. But the question remains of how a strategic customer can be distinguished from one who is non-strategic – a reference project from a business without any promotional effect. And above all: what consequences does this have for determining the price floor?

In economically rational thinking, only future associations are relevant and can justify pricing concessions (beneath the short-term price floor). If until now the provider has done good business with a customer, but it can be expected that the customer will hardly continue to "pay," then at least economically considered there is no reason for undercutting the short-term price floor. On the other hand a customer's future attractiveness is hard to determine. When customers are asked about their expectations regarding the development of business relations, they will *always* depict the common future in the most glorious colors when hoping for a price reduction. The dilemma is clear: sales prognoses are anything but certain; correspondingly they are an uncertain basis for planning. At the same time well-documented data from the past is not much help in judging a customer's future importance.

But the question arises of whether this is really so. In actuality, at the time of deciding for or against a contract, no one can know with certainty whether the relation will be successfully continued. But it is entirely conceivable that past transactions already contain signs pointing to a customer's loyalty – or suggest it is doubtful. In our view, by observing the development of contribution margins (= revenue – relevant costs) in very recent contracts with the customer, shifts in behavior can be identified signaling either a positive future or problems. From this development, the customer's sense of loyalty becomes apparent. A steady or even rising development of the contribution margins signals a stable business relationship, one very satisfying for the provider.

On the other hand, if the customer pursues a strategy of exploitation, this is automatically revealed in sinking contribution margins. The customer reallocates to his or her advantage. This reallocation carries the risk that the provider will end the relationship. The customer will only accept this risk if alternatives are available or the object of the customer-provider relationship no longer holds any great significance. It is precisely here that undercutting the usual short-term price floor needs to be viewed critically: the accepted revenue reduction is unlikely to be balanced out by future business. If, to the contrary, the customer is dependant on the relationship – the IPS^2s are important or costs of switching to another provider are high – then there will be an effort to diminish the danger of the relationship ending. In this manner, an analysis of the relationship's progress helps indicate the probability of a customer honoring the provider's accommodation in the present with increased revenue in the future. If an exploitative approach by the partner is already manifest at the time of the decision, investment in the relationship's future will be judged much more critically than in the case of a stable or even prosperous partnership.[69]

It is clear that scrutiny of recent contractual contribution margins can be very helpful in judging the loyalty of a customer. An undercutting of the short-term price floor should only be allowed when it is compensated for with loyalty and future increased revenue. It is also clear that this can only function for established customers. Having a common history is a precondition. New customers are naturally in the most difficult position when it comes to a prognosis of loyalty and strong profits. But it is true in any case that if undercutting the short-term price floor has to already be considered in initial business with a customer, then prospects for the relationship developing positively are bleak.

In sum, the discussion over the past three sections has furnished an itinerary for identifying price floors for IPS^2s and determining what influences on the approach to these price floors stem from developing or already existing relationships. Integration of direct costing with activity-based costing here offers a good possibility for mastering the mix of IPS^2's goods and services modules. It will have become clear that our discussion applies to short-term price floors. In the long term, an IPS^2 is subject to the same evaluation rules as any other company project. It has to bring its value contribution; the revenue must be higher than the costs, including an adequate rate of return, if the IPS^2 is not to be stricken from the provider's service portfolio for depleting company value.

Conclusion

This chapter has reviewed competitive challenges presently facing companies in the triad countries in order to suggest one possible path of survival. The solution analyzed here is the IPS^2: a new form of goods and services provision that can generate barriers to imitation significantly more difficult to overcome than is the case with current isolated product and service offerings. In this regard we tried to show that achieving a competitive advantage cannot and must not be sufficient for a company. Rather, the question emerges of how the benefits – above all, the entirely profane, monetary ones – are to be allocated between customer and provider. And here the rule of pure egotism prevails: as many as possible for me and only as many as necessary for the customer. Translated into the world of numbers this means that the customer's payment willingness has to be skimmed off to the highest possible degree, with one's own costs being pushed down at the same time. In order to offer analytical accompaniment, we discussed both methods for determining the customer's price ceiling and the rules for dealing with costs as the provider's price floor.

The author is very much aware that by necessity, this chapter has only constituted an overview. Each of its points needs much further discussion. Each also needs a great deal of adaptability if rather crudely developed concepts are to be applied to real companies with real marketing situations, products, competitors, and controllers. For such reasons, the above arguments should be considered guidelines for future debate rather than a ready-made solution.

5
Evolution of the Sales Force in a Global Economy

Jagdish N. Sheth
Arun Sharma

Over the past two decades, there has been a dramatic change in international markets and marketing. The major developments at play here have been the fall of communism, the establishment of the European Union, the emergence of Asian markets, and the establishment in 1995 of the World Trade Organization. These developments have led to two distinct market shifts. First, demand for products has grown, especially in emerging markets in Asia and Eastern Europe; and second, there has been increased global competition, which has led to firms examining returns on processes, reducing costs, and embracing cost-cutting technologies such as the Internet, just-in-time compilation, and automation.

The marketing process has also had to change, with a stronger focus emerging on marketing metrics and return on marketing expenditures. And this has been accompanied by an evolution of sales force strategies: the use of salespeople and communication networks to allow personalized, uninterrupted communication between the firm and its customers. The traditional sales force was organized around products. Newer approaches such as the customer-focused sales force, sales automation, and the key or global account management sales force have now emerged. The design and implementation of sales force strategies in international markets has added an increased level of complexity. Although some authors have suggested instituting shared strategies due to commonalities across international markets, our finding is that the different development and sales culture of different countries places varying demands on their sales force. We thus suggest developing strategies targeted towards these differences.

At the same time, for instance, that there has been an increase in sales automation and global key accounts in the United States, there has been a very large increase in face-to-face salespeople in India. We will thus offer a framework for classifying countries according to their economic development and sales culture. Based on this classification, we will then review the type of sales force strategies that will be successful in different international contexts.

Classification of Countries

To examine the effect of specific national contexts on sales force evolution, we have classified countries according to two dimensions: their economic

Bringing Technology to Market. Olaf Plötner and Robert E. Spekman

ISBN 978-3-527-50270-7

development and sales culture. Previous research has indicated that both dimensions affect marketing practices.[70]

A country's economic development affects basic infrastructural elements such as roads and telecommunications, as well as political elements such as the existence of elected legislative bodies and independent judicial systems. There are two reasons why infrastructure development is important for the sales force. In the first place, telecommunication density, broadband access, data-processing devices (including personal and handheld computers), and mobile phones have come to play a large role in gaining access to customers. In the second place, a country's legislative and legal infrastructure has an impact on costs of communication and compliance: these are greater when that infrastructure is weak, this being tied to processes ensuring that exchange partners do not take advantage of other partners.[71]

Sales culture is the level of acceptance of sales practices; it is enhanced through marketing institution development, itself tied to the availability of competitive marketing institutions – efficient and effective distribution and communication channels. Such development allows easy access to customer information from institutions such as consumer credit-reporting agencies and business databases by third-party vendors. Notably, because this sort of development can be decreased by domestic rules and regulations, it is not always correlated with economic development. For instance advanced countries such as Japan have low levels of distribution channel effectiveness, as in Japan these channels are long and cost is high. When marketing institution development is low and firms cannot use traditional strategies such as brand building, the role of the sales force becomes more critical.

It is now possible to divide countries according to the two-by-two matrix produced by the above classification. Developed economies and sales cultures are present in advanced countries such as the United States and Hong Kong. Firms in these countries follow leading-edge sales strategies; the countries have invested resources to develop infrastructure and have rules and regulations allowing the development of competitive marketing institutions. For example, in the United States, anti-trust and resale price maintenance laws and similar legislation have created an environment of enhanced competition. This has offered firms access to efficient and effective distribution and communication channels for marketing their products to customers. The advanced sales culture in such countries is reflected in the presence of universities providing strong sales and sales-management programs, with careers in this realm being viewed as potentially leading to advanced executive positions.

Some countries, for example Japan and Germany, are economically advanced but have an emerging sales culture. Firms in these countries follow culture-based sales strategies. Germany has traditionally had an engineering culture and selling tasks have mainly been communicative rather than persuasive in nature. In addition, restrictions on labor and trade have made it difficult for firms to develop traditional marketing strategies, so that the sales force has

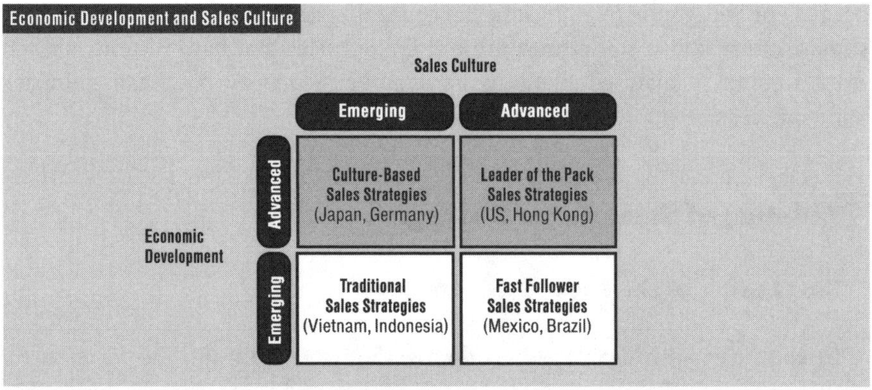

Fig. 1: Economic Development and Sales Culture

become a more important, strategic resource. In these countries, due to stringent privacy laws (Germany) or market norms (Japan), third-party data on customers is not easily available, making the sales function even more difficult.

Some developing countries reveal an inverse sales pattern to the above – they have low levels of economic development but a developed sales culture. These countries pursue a fast-follower sales strategy – in other words they quickly pick up on the sales innovations of those countries with developed economies and sales cultures. Good examples are Mexico and Brazil, which have distribution institutions such as Wal-Mart and Carrefour and have highly developed print and television media. In Brazil, Avon uses 400,000 direct salespeople, some of whom do their work in the interior region of the Amazon.

The last category in our classification includes countries with low levels of both economic development and sales culture. These are typically developing countries that have not developed efficient and effective marketing systems

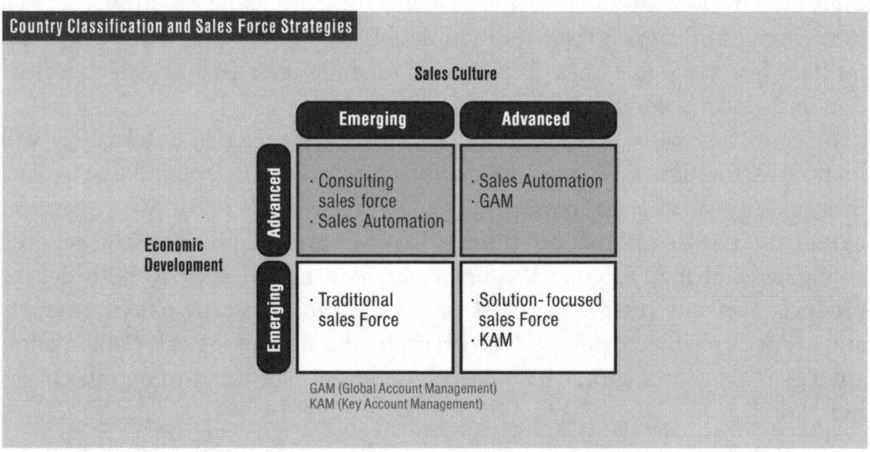

Fig. 2: Country Classification and Emerging Sales Force Strategies

because of geography (e.g., Indonesia) or legal restrictions (e.g., Vietnam). The same pattern is also sometimes present in countries that have recently shifted from a centrally planned economy to a market economy. All these countries will typically use traditional sales strategies.

Varieties of Sales Force Strategies

The "Leader of the Pack" Sales Strategy

In countries with developed economies and sales culture, the focus is on reducing the marketing costs of the sales force and increasing its reach. For such reasons, these countries have seen a decline in the traditional sales force over recent decades. In the last fifty years, sales costs in the United States have escalated from around 20 dollars per sales call in 1950 to about 200 dollars in 2000. Since it takes between three to five sales calls to close a sale, and not all calls lead to sales, the effective cost of a salesperson is about 2,000 dollars to 3,000 dollars per order. Furthermore, a survey by McGraw-Hill has suggested that 39 percent of a salesperson's time is spent face-to-face with customers, 24 percent is spent going to meetings and filling out paperwork, and 32 percent is devoted to traveling and waiting, with a final 5 percent going to customer service. Likewise, firms have frequently reported skyrocketing costs for selecting, training, and retaining sales personnel. Finally, we can observe a reduction in the effectiveness of salespeople. For example, a salesperson at IBM may be paid 150,000 dollars and may generate 2 million dollars in business. In contrast, a telemarketer at Dell may be paid 40,000 dollars and produce 4 million dollars in business. Technology has also allowed firms to better understand their cost structures. With the proliferation of so-called CRM (customer relationship management) systems and activity-based accounting, firms have begun to more effectively grasp the principles needed for a profitable sales force, as well as the need to focus on their more profitable customers while reducing costs for their less profitable ones. This has led to an increase in both sales automation and global account-management activity.

In countries with "leader of the pack" strategies, sales automation will increase dramatically. We now see manufacturers such as Dell and Cisco effectively using such automation to reduce costs and increase reach. We also expect a dramatic rise in automation in firms that can provide products and services in digitized format. Digitized information reduces the need to both deliver products and services for the marketing firm and physically obtain products and services for the customer. One example of a firm using sales automation strategy is salesforce.com, which provides and sells services through the Internet.

There are a number of reasons for the increase in automation. First, firms are trying to reduce their costs in this manner. Second, automation allows firms to provide information on their entire offerings, so that salespeople can concentrate on one or two product lines. Third, sales-automation systems allow firms to operate all of the time, in concert with their customers. Fourth, automation allows customers to deliberate on their decisions in the context of customer adaptation (this has been the case with Dell). Fifth, it reduces errors, since customers submit their requirements directly to an ordering system. Finally, technologies such as electronic data interchange allow customer computers to communicate with marketing firm computers, reducing the need for employees. Sales automation has taken many forms; these are reviewed below.

Automation of Information Access

The two sales areas that have seen the most growth over recent years are specification of product and service and information provided to customers. Almost all businesses have Internet sites providing such specifications and information to customers. These sites typically offer more detailed information than salespeople can offer. Furthermore, when the form of information to customers is very complex, the Internet can provide an effective method of information display. Internet travel sites, readily providing all the information that a customer may need, here offer a good example.

Automation of Sales Processes

CRM processes are becoming increasingly prevalent. For example, both caller identification units and voice response units can ascertain customers' needs and suggest solutions before possible contact with a salesperson is initiated. In addition, as stated some of the ordering processes have become automated due to electronic data interchange. Firms such as Wal-Mart require that their major suppliers develop systems that can receive orders through their information management systems.

Selling on the Internet

Both firms and third-party distributors are increasing their presence on the Internet, allowing them to sell products that earlier needed the intervention of salespeople. The Internet is also becoming the main avenue for both minor product adaptations associated with computers (such as those offered by Dell), and major adaptations associated with products such as the aircraft engines manufactured by General Electric.

Reverse and Surplus Auctions

Reverse auctions are becoming an attractive purchasing method. Here a single buyer opens a fixed-duration bidding event on the Internet, with multiple suppliers competing for business. Another emerging area is that of auction sites such as eBay and liquidation.com, which sell B2B surplus and used products, including products manufactured and used by firms.

Key Account and Global Account Management

In the context of globalization, many firms throughout the world have been undergoing rapid geographic expansion. In the United States, businesses with more than 500 employees now operate from an average of 54 locations each. Instead of different salespeople visiting and negotiating with each location, these businesses seek a single point of contact, reducing the emphasis on location typifying traditional sales organizations. This has led to an evolution in the nature of sales forces focusing on large accounts. The process began with national accounts, which were transformed into key accounts and then into global accounts.[72] It has been reported that 80 percent of purchasing managers are already working with suppliers who have key account programs, and that 74 percent see their contact with key account suppliers increasing for strategic reasons. As firms start competing globally, they require sales organizations that are also global.

To cater to these customer organizations, sales organizations also need to be organized like the purchasing firm. The role of the national or global account manager becomes one of managing the fulfillment of customer needs by marshalling the resources of the firm. The role of salespeople in key or global account management becomes that of a general manager marshalling internal and external resources to satisfy the needs and desires of customers. This requires new processes spanning the boundary between firm and customer. The model at work here is analogous to the consultant model, where the lead partner, marshalling global resources to match the needs of key or global customers, is in charge of sales, profitability, and customer satisfaction.

Key or global account teams need three skills. First, the managers of such teams need to integrate new and complex knowledge to understand customers' explicit and latent needs. Second, in key or global account management, firms increasingly need to assign sales and profit responsibilities to the sales organization. Third, the managers need to be more flexible in order to enhance efficiency.

These managers should be people who can become trusted advisors, who typically reveal a solution-driven focus with a relationship-manager perspective, as compared to salespeople, who demonstrate a product focus and a transactional orientation. The global account manager's tasks are non-traditional; they include coordination, key account planning, external relationship management, negotiation, and multicultural teamwork. The incentives for this sort of management need to be based on sales growth, profit growth, and customer satisfaction, with compensation being offered along the lines of a balanced scorecard. Internal marketing is critical for global account managers; this involves promoting ideas and initiatives and shifting organizational focus towards issues such as customer-oriented marketing. Some researchers have labeled this process *reconfiguration.*

Culture-Based Sales Strategy

In countries demonstrating mature economic development but low levels of sales culture, salespeople are seen as communicative agents; the focus in these countries will thus continue to be on maintaining a consultative sales force, together with introducing sales automation. But at the same time, with economic development increasing the cost of personnel, there will be more emphasis on reducing costs, which will lead to increased automation of sales functions.

The Consultative Sales Force

The consultative sales force covers activities sometimes referred to as customer-focused selling and solution selling. Traditional sales organizations have been product based. Some firms have more than one product, customers often then becoming tired of talking to multiple salespeople from the same firm, each selling different products. This has led to sales firms sending only one salesperson to the customer; the salesperson then acts more as a customer expert than as a product expert. The customer-focused sales organization has performed better than its product-focused counterpart, thus encouraging a movement away from the latter.

At the same time the product focus of both businesses and sales organizations has been moving towards a service model based on both solutions and customer satisfaction, with the product itself playing a significantly diminished role. This shift to services and solutions has become increasingly recognized as the dominant logic in marketing.[73] With firms from different locations needing sales support and solution-focused offerings requiring higher levels of such support, the emphasis on adhering to traditional business hours has also lessened. Instead, there is more congruence with the customer's operating hours; candidates for sales positions must now expect to be available at all times. At the same time, there has been a movement of customers from an ownership to an on-demand (subscription, pay-for-usage) model. Some businesses made the shift long ago (e.g., those owning power plants until the mid-nineteenth century); others are only doing so at present.

Such factors have led to an increase in customer-focused or consultative selling. Salespeople now need to understand customers' problems and come up with solutions, the customer being optimally considered a trusted partner and the sales organization responsible for its own balance sheet. Within this new consultative framework, traditional compensation based on sales may not be adequate: a profit component may well be needed, together with indicators of customer satisfaction. This process will have an impact, in turn, on the domain of internal marketing. Since sales organizations will need resources to satisfy customer needs, they will have to do a better job of obtaining resources from their own firms.

As companies become larger, there will be a need for key account and global account management teams. In addition, as firms operating in a global environment start seeking large global firms as customers, they will need to develop global accounts. A final area of growth will be in the area of sales automation, as that will allow firms to bypass existing channels of distribution. This type of successful marketing strategy has been followed by firms such as Dell, IBM, and Movado.

The Fast-Follower Sales Strategy

Countries that have low levels of general development but have a developed sales culture will use a customer-focused sales force. One chief reason for this is that the cost of salespeople is low, due to lower levels of economic development. Another reason is that since these countries have periods of shortages and surpluses, a personal relationship between buyers and sellers is particularly important. Finally, in the context of a developed sales culture, the sales force is frequently used by competitors, knowledge of the customer then becoming a crucial factor. Notably, in such countries sales automation will not provide the cost savings typical of advanced economies; customers in any event here prefer face-to-face contact to placing orders through sales automation devices.

The Traditional Sales Force

Marked by smaller firms and an absence of sales automation, countries with both low economic development and an undeveloped sales culture continue to rely on the traditional product-based sales force. Salespeople here work in designated geographical areas, customers typically interacting with them in standard business hours. The salespeople tend to be product experts, with some firms having different sales forces selling different products.

In this context, with the salesperson both serving as the firm's communication channel and participating in the customer selection process, the selling process has been designed around an *AIDA* persuasion model: a desired customer movement from attention to interest to desire to action. This model corresponds to a process involving seven selling phases: (1) prospecting, (2) the pre-approach, (3) the approach, (4) the presentation, (5) overcoming objections, (6) the close, and (7) the follow-up. The first three phases here correspond with customer selection and attention, the next two phases with interest and desire, the sixth phase with action, the final selling phase simply involving a follow-up for future sales.

These sales processes are regarded as discrete, that is, each sale constitutes a new process. In some cases, the sales force is divided into *hunters*, those salespeople who are good at locating new customers for the firm, and *farmers*, those good at maintaining sales relationships. The selection of salespeople is based on personality, most compensation being in the form of a salary and modest commission.

6
Organizing the Sales Force

Wesley J. Johnston
Brian N. Rutherford

Along with needing to understand marketing strategy and the more quantitative aspects of sales such as forecasting, sales managers must be skilled at the people-oriented tasks involved in organizing a sales force. These tasks are closely related to human resources management and an understanding of individual psychology. How well sales managers perform the tasks and organize their sales force directly affects their ability to reach their organizational objectives. This chapter offers an overview of the foundations for organizing an industrial sales force (see fig. 1). Once this foundation is provided, we will discuss the importance of recruiting, selecting, and proper training. We will then turn to issues related to compensation, incentives, and evaluation.

Organizing the Sales Force

There are many guidelines managers need to focus on to create a good organizational structure. Several widely accepted guidelines share the following factors: division of work, span of management, and change (see fig. 2). In any organization, it is important that employees can be assigned different tasks and responsibilities, a designation of authority allowing managers to realize the organization's effort in a coordinated and integrated way. If such a division of work is not present, the efficiency of the sales force will become greatly reduced. At the same time, the ability of a supervisor to manage large numbers

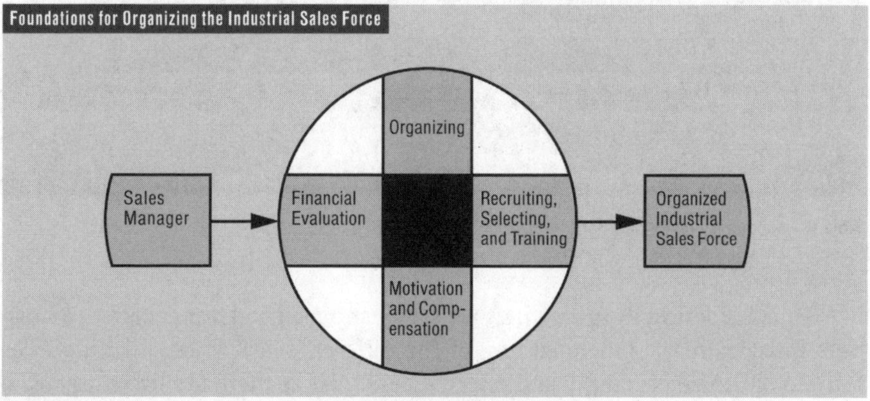

Fig. 1: Foundations for Organizing the Industrial Sales Force

Bringing Technology to Market. Olaf Plötner and Robert E. Spekman

ISBN 978-3-527-50270-7

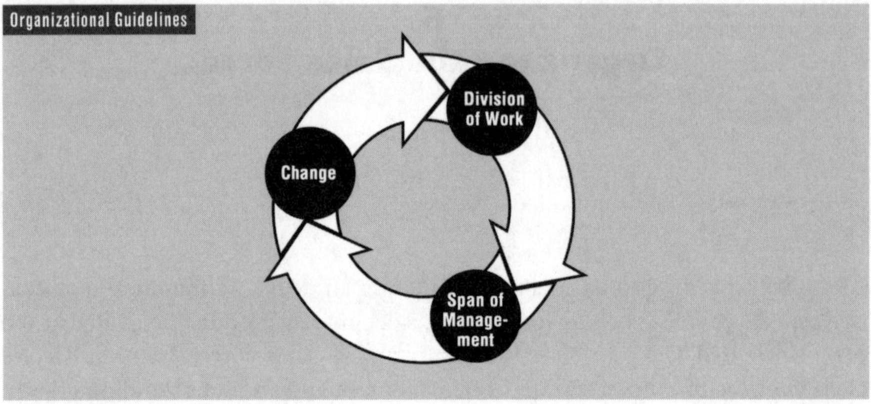

Fig. 2: Organizational Guidelines

of people is limited. Typically, it depends on the complexity of the job, the frequency of contact, the dispersion of the sales force, the extent to which the manager can delegate authority to others, and the training and capacity of the manager and the sales force. In most cases, a first-level manager can supervise more salespeople than managers higher on the corporate ladder.

We are witnessing an era of organizational transformation and dramatic change in customers' needs.[74] It is necessary that everyone working in the organization understand that such change is inevitable. In order that businesses remain viable for real and potential customers in the long term, the change needs to be carefully planned and objectives established, while short-term objectives are nevertheless realized. If a business fails to achieve a balance between the long-term and short-term objectives, its ability to survive will be reduced.

Making Changes Easy for Salespeople

It's human nature to resist change, and often it's smart business sense to resist it when you're part of a sales team. Reorganizing a sales force can disrupt deals by putting reps in unfamiliar territories, adding new products, and causing confusion about responsibilities. But a competitive team must change to compete, and if it's done correctly, reorganization can occur without any sales loss.

For a painless transition:

- **Get the message out to reps early about planned changes. Informed employees buy into reorganization strategies quickly.**
- **Don't keep reps away from accounts by saddling them with meetings and training. Instead, spread these sessions out over time.**
- **When reorganizing territories, be as gentle as possible. Don't take away hard-earned accounts from reps.**
- **After a major change, follow up with both employees and customers to make sure goals were met.**

Table 1: Making Changes Easy for Salespeople[75]
Source: McMaster: Making Changes Easy

After considering the guidelines for establishing a good organization, a business must gain an understanding of the different ways to organize its sales force to optimize efficiency and effectiveness. Just as there are basic organizational structures – e.g., the line, line and staff, and functional structures – sales

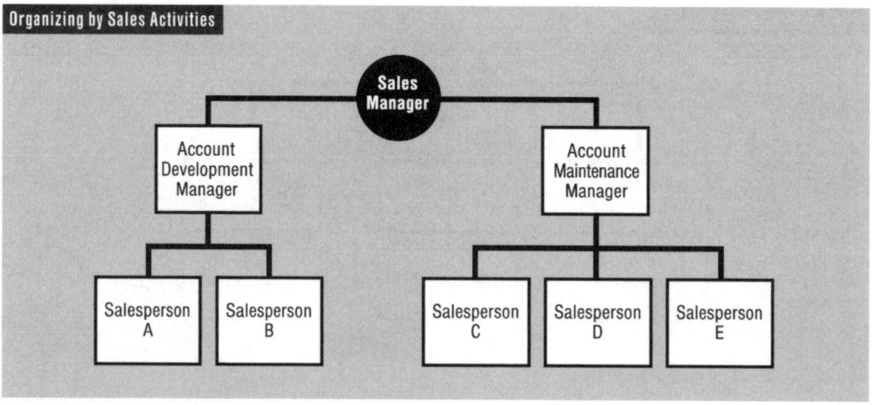

Fig. 3: Organizing by Sales Activities

managers can organize the sales force based on different attributes. Common sales force organization is by sales activities, geographic areas, products sold, and customer type.[76]

Organizing by Sales Activities

A number of companies have found it helpful to divide the sales force into two groups (see fig. 3). The first of these is the new-account development group, responsible for finding new customers; the second is the account maintenance group, which – as its name suggests – maintains an account after it has been established by the new-account development group. The most obvious advantage of this form of organization is that it allows both groups to become proficient in their responsibilities, together with offering a special emphasis on searching for and selling to new accounts. Its main disadvantage is that customers may not like having their accounts turned over to different salespeople. In addition, it does not allow the salesperson who has acquired an account to cultivate and develop a relationship with the customer. Before implementing this type of strategy, the organization should consider the consequences from both the customer's and salespeople's point of view.

Organizing by Geographic Areas

As the geographic area in which a company conducts business expands, increases in inefficiencies may prevent efficient management of the sales force from one central location. To combat geographic limitations, firms may choose to organize the sales force on a geographic base (fig. 4). In that case, all of the company's products are typically sold by the salesperson assigned to the given geographic area. As the number of products sold increase, the assigned geo-

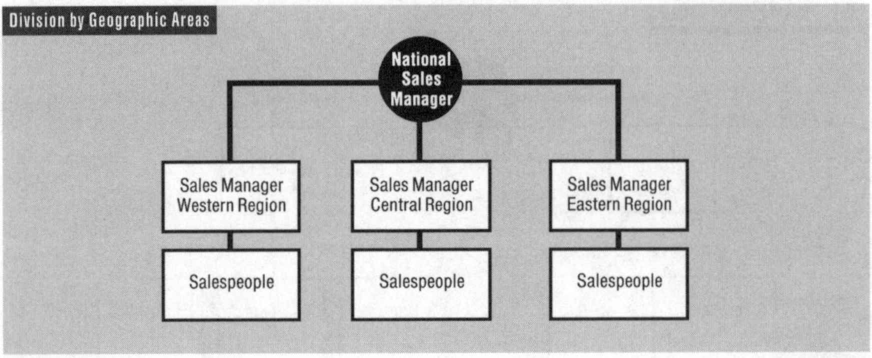

Fig. 4: Division by Geographic Areas

graphic area is reduced for the salesperson. One of the main advantages of this type of organization is the salesperson's ability to cultivate the local markets more intensively and adapt to local problems and conditions. In addition, these problems can be addressed quickly, without having to wait for a home office decision. A clear disadvantage emerges when the company's product line is heterogeneous. As the product becomes more heterogeneous, the sales force's ability to understand and retain sufficient knowledge to represent all products to customers is substantially reduced. Consequently, this organizational strategy is best for companies that have a relatively homogeneous product line and sell to customers who are geographically dispersed.

Organizing by Products

As the products an organization sells become complex or greatly varied, a product-oriented sales force may be most appropriate (fig. 5). When a company adopts such a sales force, each of the many product lines it has will receive

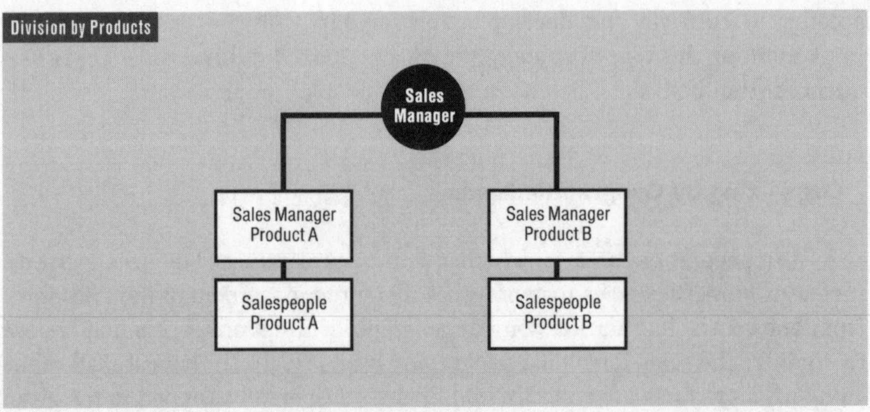

Fig. 5: Division by Products

a higher level of specialized attention. This approach allows for decentralized authority and responsibility for sales within each product line; necessary decisions can be made by those closest to the issue or problem. One of the approach's principal disadvantages is that more than one salesperson may call on the same customer because of different products being sold. This overlap may not only create problems for the organization in creating a consistent company image, but may also generate confusion and frustration in the customer base. The approach works well when the company's product line is sold through different channels of distribution and to different markets, and when the buying center in a single buying organization is made up of different people for the different product lines sold.

Organizing by Customers

Companies may find that their customer base is made up of entirely different segments. In this case, the company may need to organize the sales force based on individual customers, on industries, or on channels of distribution (see fig. 6 for an industries example). When this form of organization is chosen, consumer orientation is at the forefront, the company being oriented towards the special needs of specific customers or classes of customers. The main disadvantage of this strategy is the high possibility of overlapping territories. In addition, the sales force is burdened with a need to be familiar with the firm's entire product line.

Recruiting, Selection, and Training the Industry Sales Force

For a firm to survive, it must have the ability to generate revenue. Because the sales force is a revenue generating source, the firm needs to maintain enough sales personnel to accomplish its objectives. Because of turnover and expansion, a sales manager typically needs to continually search for new talent

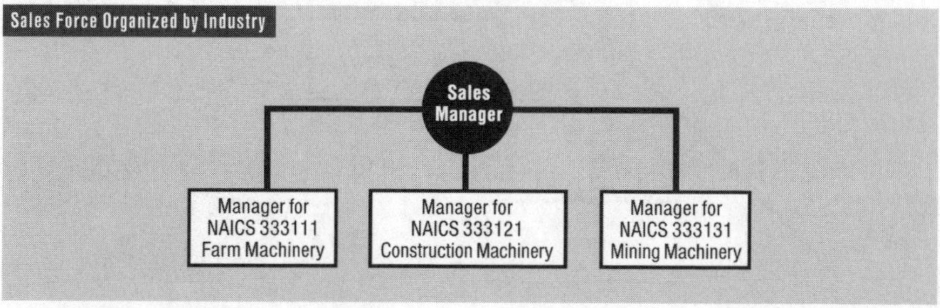

Fig. 6: Sales Force Organized by Industry

in order to maintain a sufficient number of sales personnel. While in many cases turnover happens unexpectedly and may not allow enough time for selection, recruitment, and training of a new salesperson to take over the departing salesperson's position,[77] proper planning may reduce some of the negative impact.

When working to recruit new salespeople, the sales manager needs several key documents. First he or she needs to create a job description – a written document serving as a guideline for recruiting, hiring, training, compensating, supervising, and evaluating the sales force – if one does not already exist. A proper job description depends on performance of a job analysis, the basic steps of which are summarized in fig. 7. The job description itself should include five sections. The first of these should be a planning section, describing all preparatory work required before an actual sales call can be made. Next comes a selling section, outlining the salesperson's responsibilities when making an actual sales call. The third and forth sections should contain information on responsibilities for both servicing customers and passing along information; and the final section should include any miscellaneous responsibilities not listed elsewhere. The sales manager and the salespeople should also develop a set of job specifications, detailed requirements for successfully performing the described job.

Once the job specifications have been formed and a job description exists, the sales manager should then look for salespeople. There are many potential sources for such individuals; several key sources will be reviewed below. For a start, other departments throughout the organization can provide a good

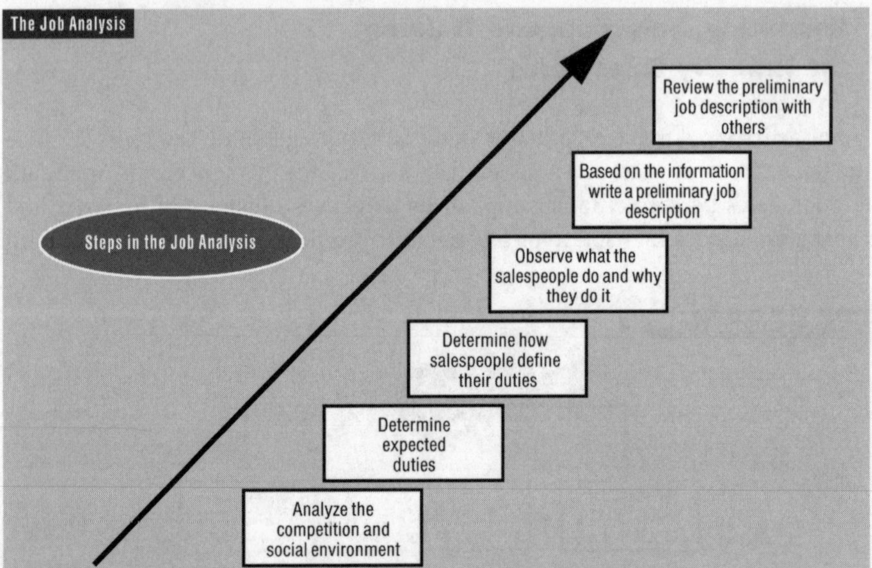

Fig. 7: The Job Analysis

source for salespeople. One advantage of using this source is that the recruit will already be familiar with company policies and may possibly know about various aspects of the product being sold. Next, the sales manager can ask current employees for recommendations for new salespeople. Third, the career services departments of various educational institutions can serve as an excellent source for salespeople. Lastly, individuals at competing firms can be recruited (see table 2). This is a controversial practice, and sales managers need to carefully consider its legal and ethical ramifications. The chief advantages of recruiting a competitor's salespeople are the knowledge of the industry they already have and the possibility of bringing existing customers to the firm. The chief disadvantages are the usual need for high levels of starting pay, a reduced tendency to stay with the firm for an extended period of time, and the above-mentioned legal and ethical factors.

Stealing Salespeople: Pros and Cons

Pros

Ramp-up time: Salespeople from competitors can start selling for your company from their first day on the job. They'll need less training and can be generating revenue faster than a rep who is new to the industry.

Staff motivation: If you're feeling any complacency on the part of your salespeople, give them a hint that you're looking at hiring a couple of reps from competitors. Nothing gets a salesperson's competitive juices flowing more.

Experience: A competitor's salesperson will come to your company with years of industry experience, knowledge about the market's products, and a wealth of customer contacts.

Cons

Clash of cultures: Salespeople from rival companies may find it hard to adjust their selling styles. If they're used to a different kind of communication system or a different way to approach customers, be wary of hiring them.

Non-compete clauses: While the enforcement of non-compete contracts varies from state to state, managers should check with a lawyer before getting too far into the recruiting process.

Credibility with customers: If salespeople are calling on customers saying one product is the best on one day and peddling the greatness of a competing product on another day, their honesty and integrity–as well as that of your company–will be in jeopardy.

Table 2: Stealing Salespeople: Pros and Cons[78]
Source: Chang et al: Should You Raid Your Rival's Sales Force?

When selecting the sales force, hiring the right individual for the position is one of the most critical concerns for sales managers. Clearly, hiring the wrong salesperson may well have a negative impact on sales productivity. If the person leaves the position, the costs involved in training will be lost.[79] With that said, the primary objective of any hiring process should be to choose individuals with the greatest potential for selling success and who seem likely to remain with the organization over a long period. Fig. 8 shows the steps in the selection process.[80]

During the primary screening process the obviously unqualified candidates should be eliminated. One useful tool in this process is the application form. During the second phase, the initial interview, candidates should be considered with regard to their dress and appearance, personality, interactive skills, verbal ability, motivation, interest in the firm, and other historical and factual infor-

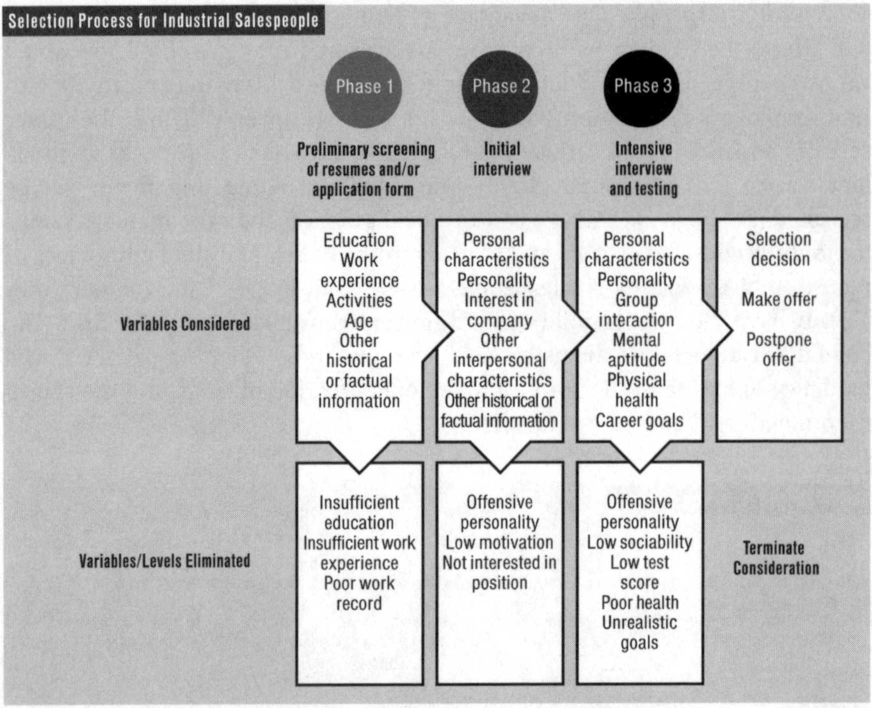

Fig. 8: Selection Process for Industrial Salespeople

mation. While there are different types of interviews, two of the most widely used are nondirective and patterned interviews. Nondirective interviews allow the interviewer maximum discretion in determining what will be discussed. In this case, the interviewer only asks enough questions to maintain the discussion. In contrast, patterned interviews are highly structured, with the interviewer following a specific set of questions and recording the answers. The results of patterned interviews can be compared across candidates. In the next phase, intensive interviews and testing are conducted, with personality, group interaction, sociability, mental aptitude, physical health, and career goals being considered. Finally, a decision is reached to make an offer or not do so.

After the salesperson or salespeople have been selected, new recruits should engage in a training program. Its contents should include knowledge of the product, the company, and the market, selling techniques, customer motivation, and non-selling activities (see table 3). Product knowledge should include information about how the product is made and used and what the competing products in the market are. Information about both the product's features and its benefits should also be conveyed. Company knowledge should include information about what the company does, what it stands for, and its basic structure and procedures. Market knowledge should include general information about business conditions and specific information on the industry in question. As indicated, the new recruit should also receive information regarding both different aspects of the selling process and what motivates and does not motivate customers. Finally, information should be provided on non-selling activities such as customer service, generating sales inquiries, time and territory management, and paperwork.

After the initial training program has been conducted and the sales force has been in the field for three to six months, continuing training should be implemented. This is an extension of the sales force's initial training. Considering that the average cost of a single industrial sales call was in excess of 262 dollars according to a 2000 study, continuing training of the sales force is important to maintain and increase its performance.[82] Continuing training is needed because it helps reorient the sales force toward acceptable selling procedures, has the potential to provide it with new knowledge, and helps it both obtain new skills and develop new attitudes. A common setting for continuing training is the sales meeting.

The Importance of Sales Training

What is Necessary	What is Happening
The V.P. responsible for the 500% sales increase of a division of Fort Howard Corp. attributed the cause to sales training (Keenan, 1995). There is a positive relationship between sales training expenditures and share price (Hall, 2004). A toothbrush manufacture credits sales training for a $30 million sales increase (Chang, 2003).	63 percent of sales managers indicated that they neglected personal skills development and indicated that they prioritized revenue over profit (Gilbert, April 2004). Only 47% of surveyed sales managers indicated that sales training was very valuable (Galea and Wiens, 2002). 37% of industrial firms surveyed did not train their salespeople (Puri, 1991). In another survey by Learning Inernational, 43% of industrial sales managers indicated that they did not train their salespeople (anonymous, 1989).
The right sales training can produce significant increases in sales (Klein, 1997; Zoltners et al., 2001). A Cartnell survey indicated that the costs of sales training for industrial products ($9,893) and services ($9,060) are higher than the $7,079 average sales training cost (anonymous, 199a).	60% of firms do not formally evaluate the financial impact of sales training because of perceived difficulties and expense (Galea and Wiens, 2002). When introducing sales training programs, these problems are faced: top management is not dedicated, program is not adequately funded, salespeople are apathetic, salespeople resent intrusion on time, and salespeople resist changes (anonymous, 1990a).

Table 3: The Importance of Sales Training[81]
Source: Pelham: Sales Force Training

Motivation and Compensation

One of the most frustrating jobs sales managers have is motivating their salespeople – a crucial task in light of the rejection and frustration that a sales force sometimes faces. One source of frustration for managers in this respect is a lack of experience and training in motivation. Tables 4 and 5 offer suggestions for helping to motivate salespeople.

Individual Salesperson	The Sales Job	The Organizational Environment
Recognize and become sensitive to the unique needs of each salesperson	Provide sufficient product knowledge	Take active role in developing motivational processes
Give praise and recognition	Communicate the professional status that the salesperson's role connotes	Let the salespeople know their suggestions are welcome
Once financial security is met, appeal to other needs	Help each salesperson set reasonable goals and design plans to attain those goals.	Maintain an environment where trust, good faith, and fairness are openly practiced
Explain job requirements and expectations	Provide sales positions with challenge, authority, and freedom	Communicate that the salespeople are mature, professional individuals involved in significant, meritorious work
Train salespeople to be professionals	Recognize the frequent adversities encountered by the salesperson	Communicate the relationship between performance and recognition
Provide rewards that are simple, immediate, frequent, and related to a special act	Prevent job dissatisfaction by providing a fair compensation plan, constructive supervision, fringe benefits, and job security	
Be aware of the value of personal visits, personal phone calls and letters, and intimate sales meetings		

Table 4: Applied Motivation

Expert Tips on What Not to Do
Don't confuse incentives with recognition. Incentives are designed to motivate changes in behaviour. Recognition relates to rewarding those changes in behaviour, if and when achieved.
Don't create goals in a vacuum. Work with individuals and team to create clearly defined, measurable goals; make sure everyone is on board with expectations.
Don't target your efforts to just your top performers. Motivate the middle. Structure your incentive program so that it is open-ended and on an individual basis so that everyone has a chance to win.
Don't forget to communicate. Track progress. Let participants know how they're doing and how their competition is doing.

Table 5: Expert Tips on What Not to Do[83]
Source: Chang et al: Happy Sales Force, Happy Returns

Monetary compensation is a motivator for improving performance up to a certain level. (In other words, money is a lower-order factor along with company policies, relationships with co-workers and supervisors, and general work conditions.) While monetary compensation is only a motivator up to a certain level, it is needed or the sales force will become dissatisfied. When a sales manager considers compensating the sales force, several key objectives must be met to provide a strong compensation plan. From the salesperson's viewpoint, the plan needs to be equitable, stable, and understandable, and it should also provide incentive. If any of these factors are missing, the sales force will experience reduced levels of motivation. From the company's viewpoint, the compensation plan needs to: (1) attract and retain desirable salespeople, (2) encourage specific activities, (3) reward outstanding performances, (4) provide the optimum balance between costs and results, (5) develop long-term relationships with customers, and (6) provide easy administration.

Basic compensation plans can be centered on salary, commission, or bonus, or be a combination of these factors (see fig. 9). Salary is a fixed sum of money paid to a salesperson for efforts expended over a set period of time. In general, salary is paid on a weekly, bimonthly, or monthly basis. Advantages of a salary include the following: (1) the salesperson is assured of a steady income, (2) more loyalty to the company is developed, (3) the salesperson is more flexible when it comes to accepting non-selling activities, (4) there is more control over the sales force, and (5) salaried payment is easy to administer. The disadvantages include the following: (1) such payment does not offer financial incentive to perform at a higher level, (2) it can increase overall selling costs, (3) the least productive salespeople are traditionally overpaid while the most productive salespeople are underpaid, and (4) it tends to produce adequate but not superior performance.

Commission-based compensation programs pay the sales force based on some unit of performance and a rate. The rate is the amount paid per unit. For example, a rate of fifteen percent of sales price is paid to the salesperson. Given that a salesperson sells 100,000 dollars worth of merchandise in a given period, the compensation would be 15,000 dollars. Commission-based compensation may or may not involve the salespeople paying their own expenses.

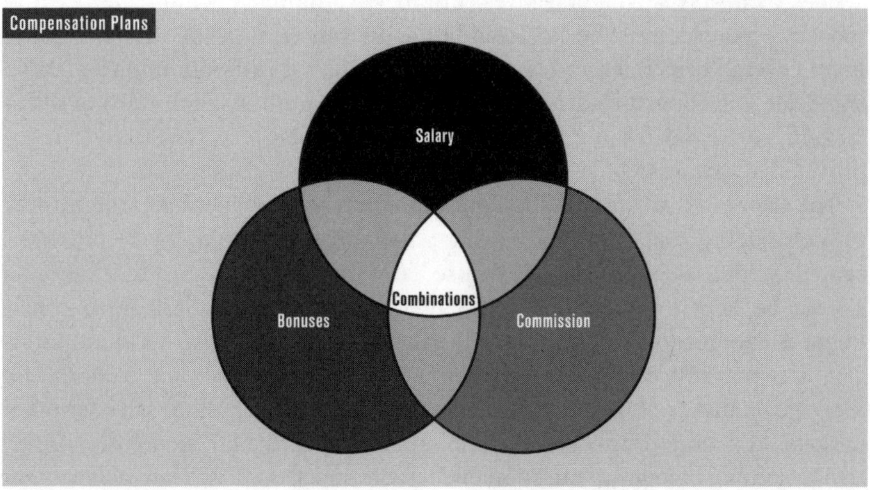

Fig. 9: Compensation Plans

Typically, the commission rate will be higher when these expenses are indeed not reimbursed. Among the main advantages of commission-based compensation are the following: (1) compensation is based on performance, (2) a commission plan is easy to compute and understand, (3) there are little or no costs when there are no sales, and (4) large sums of capital do not have to be tied up in salaries. Disadvantages include: (1) in most cases the principal focus is placed on obtaining more sales as opposed to building strong relationships with customers, (2) salespeople are not as loyal to the selling firm, (3) salespeople will often neglect non-selling activities, (4) when sales decrease or are in a slow period, sales force turnover increases, (5) salespeople have less security due to fluctuations in income, (6) switching salespeople to different locations or changing aspects of their location can create difficulties due to limited flexibility, and (7) salespeople are harder to recruit.

Bonuses are lump-sum payments for meeting or exceeding quotas or some form of achievement. For example, when salespeople reach or exceed a pre-specified performance level, they are typically rewarded with a sales-quota-bonus, which itself is aimed at enhancing their future performance.[84] Bonuses can be paid on an individual basis or to an entire group. The chief advantage of this form of compensation is its flexibility, and its chief disadvantage is tied to the potential of sales managers to use it to compensate the sales force for having to cope with unrealistic objectives.

A fourth form of compensation involves a combination of the previous three forms. The main advantage of a combination plan is its potential to merge the advantages of the others; its main disadvantage the complexity of implementation and the high chance of the sales force misunderstanding the plan.

A sales contest can be used as a form of short-term motivation. Industrial salespeople typically prefer longer sales contests (e.g., those lasting six months) to shorter ones (e.g., those lasting one to three months).[85] When planning for a sales contest, realistic objectives, in other words those within the reach of most salespeople, must be set. In addition, the contest must be tied to the company's overall principles and objectives. Likewise, the program must be understandable and communicated to the sales force, communication starting with a kick-off party and frequent updates regarding where everyone stands being provided afterwards.

The sales contest's payoff should be based on accomplishment and must be of high quality, that is, exciting to the sales force; it should not be misrepresented. While sales contests can be used as a form of short-term motivation, it should be kept in mind that unintended short-term and long-term consequences can ensue. For example, any sales systems with short-term emphasis can place pressure on the sales force to meet short-term goals and compromise those goals that are long-term. This process can include pulling sales for other periods, lessening cooperation, overemphasizing particular products, exaggerating claims, accepting questionable credit risks, and setting aside other responsibilities.[86]

Financial Evaluation

Salesperson evaluation is a complex task because of the wide range of factors influencing a salesperson's activities. Different firms require different tasks or procedures from their sales forces, and these will require different methods and approaches in evaluating performance.[87] An individual salesperson's performance needs to be measured against and compared with desired performance. When considering the financial indicators at play here, several key areas need to be addressed: financial quotas and the sales force's expenses need to be reviewed, followed by a general analysis and discussion of the salesperson's strengths and deficits.

Allowing a uniform comparison of individuals throughout the organization,[88] sales quotas have three basic uses: they provide incentive, a basis for compensation, and a method to evaluate performance. When developing a quota, the sales manager needs to make sure it is fair, sound, understandable, accurate, and reachable. Because quota plans impose objectives on each salesperson and offer the prospect of rewards, a failure by managers to provide equal treatment to all on the sales force may lead to job dissatisfaction, low morale, and turnover.[89] There are many different types of quotas; five of these are reviewed below.

Sales quotas: When sales objectives are assigned to specific sales units such as regions, districts, or a salesperson's locality, these quotas are usually expressed in terms of sales volume.[90] In general, sales quotas are easy to implement and administer. One disadvantage of a sales quota is a focus on volume rather than profitability. This type of quota is most appropriate for fast-growing markets where margins have not been cut drastically.

Expense quotas: When products have reached a state of maturity, the cost of obtaining additional sales is prohibitive. For a company to increase or maintain profit margins, it must decrease its expenses. In this case, an expense quota may be implemented. A salesperson who meets or has results below his or her expense quota may receive a bonus or other form of compensation.

Profit quotas: When gaining additional sales volume is cost prohibitive, a profit quota may be called for. Profit quotas are aimed at increasing the sales of more profitable products or selling products to more profitable customers. The chief disadvantage of this strategy is that the sales force may focus on highly profitable customers or products and neglect those which are less so.

Activity quotas: When certain activities are important for increasing sales volume, activity quotas may be put in place. For instance, a sales manager may find it is necessary to require the salesperson to call on a certain number of new customers per week. Another example is requiring the sales force to spend a certain amount of time setting up displays.

Combination quotas: Often a combination of the above quotas may serve the company best. For example, salespeople selling products in a mature phase of product life may be assigned both an expense and profit quota. Companies can also use some combination of profit quota and activity quota.

Several key factors need to be considered in deciding how to handle expenses incurred by the sales force. One such factor is the cost of generating sales. If expenses are too high, the company may not make as much money as it would like to or may even lose money. Methods for controlling expenses can range from having the sales force pay all of its own expenses to having the company do so. Several expense-payment plans are reviewed below.

When *the salesperson pays all his or her expenses*, he or she will typically receive a higher rate of commission. One advantage of this type of payment plan for the organization is the lack of administrative work involved. One disadvantage is inequity from region to region.

The *expense-quota plan* involves determining the optimum ratio of sales expenses and revenues. Salespeople who exceed this ratio are responsible for payment of the expense; those with results below the ratio are entitled to a bonus.

With the *flat allowance*, a salesperson receives a given amount of money each month to pay for expenses. The only limit is on the total amount spent, which allows the salesperson to make tradeoffs. In this case, the company can build in flexibility by giving different amounts to different regions.

Limited payment plans set limits for given areas. For example, the company may set a limit of 100 dollars a day for lodging and 100 dollars a day for food. The main problem with this type of plan is that it encourages cheating when expenses are high in one area and low in another area.

Unlimited payment plans allow salespeople to be reimbursed for all legitimate expenses incurred; they are expected to turn in itemized lists and receipts for these expenses. The biggest advantage of this plan is flexibility. The biggest disadvantage is the incurrence of unnecessary expenses.

The sales manager is responsible for evaluating the sales force – an important process for ensuring that goals are being met and costs are in line with expense quotas and the budget. A five-step process to properly evaluate the sales force is outlined in fig. 10.

Sales managers should first develop a job description if they have not done so already (see above sections on job analysis and job description). Once this step is complete, managers must select the criteria for evaluation and develop specific measures of performance. For example, quantitative indicators such as the ones listed in table 6 can be selected. These indicators tend to be objective and are easier than qualitative indicators to standardize across the sales force. Although more subjective in nature, qualitative measures can be used to focus on certain of the sales force's competencies – for example, product knowledge; relations with co-workers, supervisors, and customers; planning skills; and attitude.

Next, managers should develop benchmarks and evaluate actual performance. Both can be established and measured using a technique called *data-envelopment analysis*, in which a salesperson's performance (on any of the above evaluation criteria) can be compared with the best performer, yielding a mea-

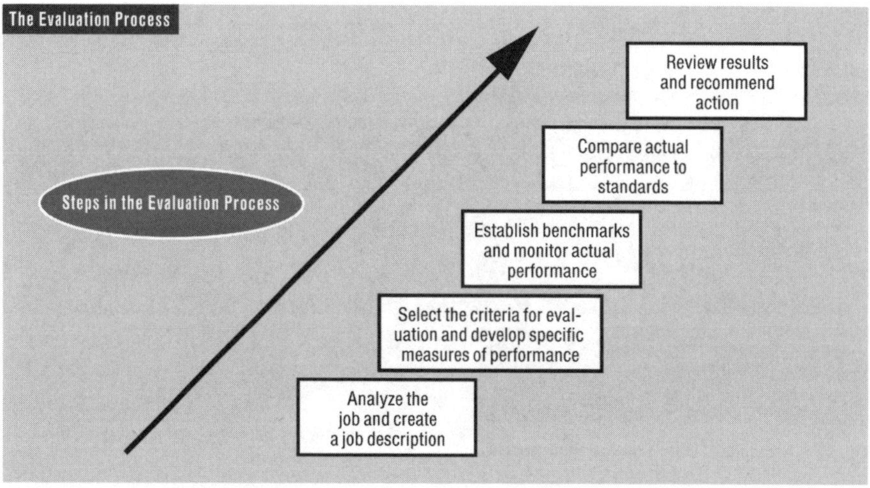

Fig. 10: The Evaluation Process

sure of relative performance. One of the technique's advantages is that managers can compare multiple inputs and outputs, forming a single measure of performance referred to as the *relative efficiency*. The analysis will compute this efficiency rating based on the difference between the value measured (actual inputs and actual outputs) and the value if the salesperson were 100 percent

Sales Force Evaluation Criteria

Sales Sales volume Sales volume by product or product line Sales volume by customer or customer type Sales volume per order Sales volume by outlet or outlet type Sales volume per call Market share	**Orders** Number of orders taken Number of orders cancelled Order per call ratio Strike rate (orders/quotations) Average order value Average profit contribution per order Value of orders to value of quotations ratio
Profit Gross profit generated Net profit generated Gross profit as a percentage of sales volume Net profit as a percentage of sales volume Return on investment Profit per call ratio	**Accounts** Number of new accounts gained Number of accounts lost Amount of new account sales Number of accounts on which payment is overdue Proportion/number of accounts buying full product line
Calls Number of calls per period Number of calls per customer or customer type Calls on potential new accounts Calls on existing accounts Prospective success ratio	**Expenses** Sales expense to sales volume ratio Average cost per call
Other Input Criteria Number of required reports sent in Number of demonstrations conducted Number of service calls made Number of letters/telephone calls to prospects	**Other Output Criteria** Number of customer complaints

Table 6: Sales Force Evaluation Criteria[91]
Source: Jobbar et al: Sales Force Evaluation Practice

Example of Data Envelopment Analysis				
Efficiency = .085 for Salesperson 22				
Variable Type	**Variable Name**	**Value Measured**	**Value if Slack**	**Efficient**
Output	Percent of Quota Attained	100	120	20
Output	Supervisor Evaluation	5	5	0
Output	Sales Volume ($)	45,000	50,500	5,500
Input	Sales Training	5	5	0
Input	Salary	20,000	18,000	2,000
Input	Management Ratio	3	2	1
Input	Territory Potential	60,500	50,000	10,500

Reference Set	Influence
Salesperson 7	0.49
Salesperson 20	0.43
Salesperson 45	0.08

Fig. 11: Example of Data-Envelopment Analysis

efficient: based on this difference, the level of slack can be computed for each indicator, and based on the slack, it is possible to indicate where levels of inputs and outputs can be improved. Boles et al. have applied data-envelopment analysis techniques to a firm's regional sales force; fig. 11 offers an example of their research.[92]

While benchmarks can be established to measure performance, not all salespeople will attain near-100 percent efficiency. In certain cases, a performance standard may thus be set below that level, managers then needing to clearly state what the standard is and why this has been done. Based on the benchmark and chosen performance standards, sales managers will review the sales force's performance and provide direction to encourage any improvement needed. The evaluation process also serves as a basis for providing financial incentives; it can be linked, for example, with a salesperson's bonus or commission.

Conclusion

This chapter has reviewed a basic set of guidelines for organizing the sales force. While these guidelines provide a starting point for such organization, there are other complex components of sales force management such as legal and ethical issues, organizing the international sales force, sales force automation, customer relations management, forecasting, and time management. Sales managers must understand that all these components need to work together to form the organized sales force. While certain components may be more important than others, if all are not handled properly, the effectiveness and efficiency of the sales force will be greatly reduced.

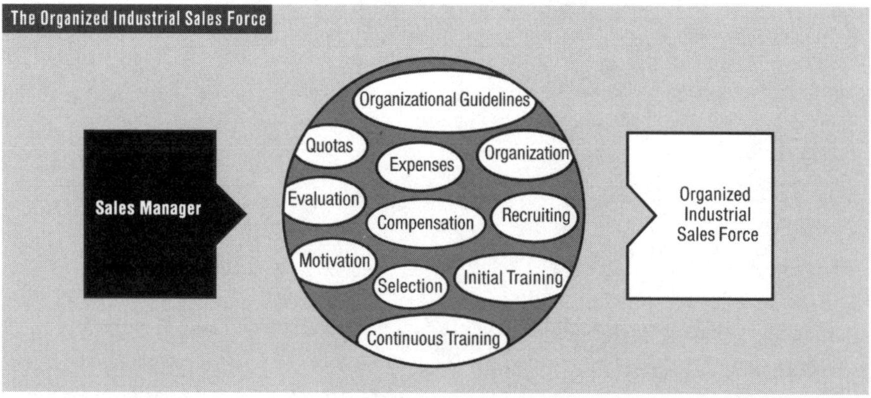

Fig. 12: The Organized Industrial Sales Force

7
Adjusting Corporate Customer Communication

Olaf Plötner
Frank Jacob

Introduction

From the 1960s onwards, marketing analysts have tended to focus on four areas perceived as at work in the marketing process: product design, pricing, distribution, and marketing communication. The latter area has itself been broken down into various marketing instruments: advertising, sales promotion, trade shows, exhibitions, public relations, personal selling, and more recently, Internet promotion and customer conferences. Over the same decades, the literature on marketing communication has focused strongly on the consumer goods market;[93] B2B markets have been broadly neglected, and when treated, only in regard to their differences from their consumer goods counterpart. In this context, B2B marketing communication has itself been approached in terms of instruments that typify the consumer goods market.[94]

The same procedure is apparent in relation to consumer goods and capital goods – a highly questionable procedure in the framework of communication activities that have considerable importance for selling complex technical products or customized services in B2B areas. For example:

(1) An energy provider issues an RFP (request for proposal) for the construction of a new power plant. An interested construction company will have to provide detailed information on its ability to fulfill the specific demands of this particular customer.
(2) A telecom provider has decided to install a new billing system. A supplier of such systems is informed of the required specifications. The supplier suggests equipping the billing system with extra, previously unknown functions.
(3) When performing routine repairs, a repairman discusses the customer's future need for spare parts.

These three examples refer to acts of communication between customer and supplier; they are directly related to the selling of B2B products and services. Still, none of them has a place in the standard model of marketing communication. Since such a list might be continued endlessly, it seems useful to develop a special approach to B2B marketing communication, one doing justice to the special situations involved in selling within this market. We believe that a suitable term for this approach is corporate customer communication (henceforth CCC).

Bringing Technology to Market. Olaf Plötner and Robert E. Spekman

ISBN 978-3-527-50270-7

Buying Processes within B2B Markets

In order to more closely consider this new approach, we first need to analyze CCC objectives, which means examining *what* a company wants to achieve with marketing communication (the objectives) and not *how* it wants to achieve these objectives (the instruments). Objectives are thus more important than instruments.

The main objective of every company is to sell its products and services profitably. Whether this objective is accomplished depends on the potential customer's decision. Unlike decisions in the consumer goods market, those about investment goods are usually the outcome of a long and complex process. Since a supplier conducts CCC activities to influence this decision-making process, it seems an appropriate starting point for a closer look at CCC. Defining CCC objectives is not complicated if we understand the decision-making process and its crucial elements, have a good overview of the supplier's opportunities to influence the process, and are familiar with the customer's need for information.

Even before making a decision about purchasing an IT-system, many people have an opinion or attitude regarding, say, the IBM brand; this predisposition could turn out to be an influential factor in any eventual buying decision. The purchasing behavior of a customer, in other words, is determined long before the actual decision process begins. Let us assume that a company has defined a problem in its production process and that it would like to solve it with a complex solution provided by an external supplier. Normally, this implies the presence of several people to consciously influence the buying process or at least try to do so.[95] The company will then begin to specify the requirements and document the results. Next, inquiries will be made about potential suppliers for the solution with the use of various information sources. The customer-firm will develop preferences on the basis of the information delivered and consequently begin a decision-making process.[96]

At the end of this phase, the customer firm will decide in favor of the supplier offering what it views as the best cost-benefit ratio. Arriving at this decision is a process stamped by uncertainty, hence very open to the influence of emotional as opposed to rational (measurable or quantifiable) factors.[97]

Once the purchasing company has selected a supplier, the buying process enters the next phase. In the case of standardized goods or services, this involves little more than signing a contract followed by delivery and installation. But in the case of goods and services of higher complexity, customers require solutions customized to their needs. This implies that the customer has to be involved in product development, especially when the purchase consists to a large extent of services.[98] For example, when buying and implementing a new software system, a customer needs to inform the supplier about the system's internal processes, provide an overview of the current IT system, and train employees on the new system. Many aspects of this process of supplier-customer cooperation take place after the actual buying decision has been made and the purchasing company has modified elements of the previously agreed-upon solution.

The cooperation between supplier and customer by no means ends with completion and delivery of the product. To the contrary, it most likely has just begun. For instance when a company decides to buy a certain type of software, it more or less commits itself to buying future upgrades as well. These sorts of delayed purchase commitments can also be observed with regards to goods such as power plants, trucks, and turning centers. To a large extent, this is due to the need for maintenance and repair. These follow-up services do not necessarily have to be performed by the original supplier. However, for reasons such as the potential loss of warrantee rights and the transaction costs saved by not having to familiarize external suppliers with a firm's systems, this is usually the case. Naturally, similar dependencies between individual transactions also exist in service-based sectors. For instance a company that has decided to hire an advertising agency for an image campaign will generally employ the same agency to develop its product advertisements in a newspaper. Such dependencies already underscore the need for buying processes not to be seen as isolated exchange activities, unrelated to past or future sales. After-sale in this context mostly also means pre-sale. The recent literature on customer relations management in B2B markets approaches the dynamic at work here as important enough to constitute a separate discipline.[99]

The Supplier's Communication Objectives

Communication during the Pre-Sales Phase

From the supplier's perspective, the communication activities that need to begin before the actual start of the procurement process can be termed the *pre-sales* phase. In this phase, the supplier's main objective is to establish contact with a targeted customer segment.

The attention gained in this phase provides the basic willingness to absorb and process further information. This does not, however, include in-depth information or information with a high level of complexity, as potential customers do not yet have sufficient interest for this. At the same time, they may remember a fragment of text, a symbol, or a name. There are many places where such information can be encountered and just as many channels through which it can be conveyed; a newspaper advertisement is just as likely as an advertisement on a local bus or everyday conversation on the work floor.

Any information a company sends to potential customers will not seem so new and unfamiliar if they have familiarized themselves with a supplier's name. If such familiarity involves a positive attitude towards the name or brand, the supplier has reached the major communication objective of the pre-sales phase, namely to create "presence." Many concepts from the realm of brand management can be applied practically in pursuit of this goal. Although primarily developed for the procurement of consumer goods, these concepts have sparked increased interest in B2B markets over recent years.[100]

Communication during the Sales Phase

In general, a supplier will either be contacted directly by a company or will have heard about its intent of purchase through a third party. Suppliers, in other words, not only need to know how to convey information (as in the pre-sales phase of communication), but also how to persuade others to share the information they own.

Once a supplier is informed about a potential customer's needs, the next step will be to ascertain the precise needs. The more accurate the information, the higher the certainty of having found the right partner for the job. Here, the need for reciprocity of information becomes clear. Not only do suppliers need to convey information concerning themselves and their available services, but they also need the communication skills to persuade others to provide required information. Furthermore, the information must be as accurate and complete as possible and delivered at the right time.

Obtaining relevant information about a potential customer before the competition can be a decisive advantage, offering the supplier both more time to create a favorable image and the opportunity to influence needs perceived by the customer. To these ends, the supplier-firm will emphasize the importance of factors where it has a competitive advantage. Communication activities working towards this goal may be best expressed by one rule: customer specifications known best are the ones written by the supplier.

Although the first aim of a supplier in the sales phase is to define the customers' needs, these are not necessarily the same as a company's needs as a whole. The people involved in making a buying decision often have very personal goals, needs, and fears. The supplier's communication activities should thus be oriented towards obtaining as much information as possible, all of it put together like a puzzle for the sake of a larger picture. Here again, the art of communication involves obtaining rather than delivering information: understanding correctly is more important than talking smoothly.

This situation changes when a customer's needs have been defined and understood by the supplier. Communication will now be aimed at the advantages a particular solution has to offer. More precisely, the supplier will have to convince the customer that the solution offers a better cost-benefit ratio than solutions of competitors. We should bear in mind, though, that suppliers cannot demonstrate the accuracy of their statements during the sales phase. For example, how high the future repair and maintenance costs of a new assembly line will be is uncertain, as is the amount of time employees will need to work with new software or how often a new machine will break down in coming years. A supplier will try to provide satisfactory answers to these sorts of questions since, in the end, a decision will be made on the basis of promises given. Consequently, the aim is not only to offer customers the best cost-benefit ratio but also to convince them that things promised will become true after the sale.

By taking a closer look at the process of customer opinion formation, we can identify two central aspects: First, it is important that the customers come to believe that the supplier has the skills required to provide the promised solution. Highly complex solutions often demand very specific skills, not only of a technical but also of a financial and organizational nature. Second, the customer needs to be convinced that the supplier actually intends to keep the promises that have been made.[101] The supplier likewise needs to make clear that there will be no reduction of activities after the sale, even when, for instance, an important assignment for another customer has to be seen to. This implies that the supplier-firm has to strengthen its own image as a trustworthy business partner. Conveying trustworthiness and skill is the second basic objective of communication in the sales phase.

Customers who are familiar with complex technical solutions can make better estimates of suppliers' skill levels. In order to judge the quality of the supplier, the customer will thus have to gain some competency in the field for which a solution is being purchased. In the absence of such competency the supplier will have to fill the gaps, with the supplied information being, however, adjusted to the customer's interests and level of knowledge. Often suppliers unnecessarily enter into details that customers find uninteresting or cannot understand.

Those suppliers that manage to convince potential customers of both the quality of their solutions and their own trustworthiness have accomplished a great deal. If the solution seems attractive enough, the supplier emerges as at least a potential business partner. Companies purchasing complex solutions on B2B markets often prepare shortlists including two to four potential suppliers. Detailed negotiations with these companies are scheduled in order to discuss all aspects of the potential sale. General clauses and promises from past conversations are turned into narrowly defined concessions, delivery dates, payment conditions, rules of guarantee, and so forth. After detailed negotiations, the agreements on these terms are usually documented in contracts. Therefore, the main objective of a supplier's communication in this phase is to commit the potential customer to purchasing the solution and negotiating conditions of sale acceptable to both sides.

This means the supplier has to make promises that will favorably influence the customer's decision and provide a decisive advantage over the remaining competitors. This must be done while bearing in mind both that the sale needs to be profitable and that all promises need to be kept. This clearly represents a dilemma, as the supplier must focus on obtaining the revenues involved in the project and yet needs to make substantial concessions in order to do so. In these negotiations, the potential customer will in fact often try to create an atmosphere that puts the supplier under pressure, thus extracting unintended concessions. This pressure can be generated in many ways, but there also are means for coping with it.

Suppliers who allow themselves to be lured into concessions they cannot keep run the risk of future losses in the form of penalties, compensation payments, and so forth. Furthermore, in the long run they will be subject to customer discontent, in other words to the loss of painstakingly accumulated trust. The discontent may be conveyed to other companies, thus causing damage moving beyond this particular business relationship.

Unrealized expectations can prompt customer dissatisfaction. Suppliers thus have to work just as hard at avoiding unrealistic expectations before the sale as on satisfying realistic expectations after the sale. This "expectation management" not only applies to the suppliers' side of the contract; rather, when necessary suppliers need to clearly explain to customers that they will also have to contribute to a given solution's development process.

Communication during the Post-Sales Phase

The supplier's objective should thus be to negotiate conditions causing the customer to decide in its favor, the deal to remain profitable, and a successful product to be developed. Especially the last point plays a crucial role during the post-sales phase, which begins when a potential customer has entered a contractual commitment to purchasing from a supplier. This turns a potential customer into a real customer. As indicated, communication in this phase is particularly important when it concerns a solution that is highly customer-specific. Whereas suppliers of crude oil, aluminium bars, or paperclips face few onerous communication challenges, suppliers of complex products and services depend heavily on their customers in the solution's development. For this reason, a successful communication policy in the post-sales phase is very important for the overall success of the sale.

In the early post-sales phase of complex products and services, a solution's definition becomes increasingly concrete. Specifications will here almost certainly determine how items are to be dealt with. Documents provide an overview of the agreements regarding the solution, the deadlines, and the division of tasks between customer and supplier. This is especially important when a customer-firm purchasing a complex solution modifies its demands after the buying decision has been made. Complying with these new demands often involves extra costs for the supplier and lowers the profit margin. The supplier should by no means block these demands; it has to be made clear, though, that they require a modified solution and as such exceed the agreement. In most cases additional negotiations will have to take place in order to comply with modified demands. Over the past years, many companies have devoted substantial attention to these "post-sale negotiations." The discipline of "claim management" is mainly concerned with the question of how suppliers or customers may best direct the outcome of communication processes to their own advantage.

Good preparation is a necessary condition for successful cooperation, but not a sufficient one. Even well-prepared cooperation can lead to friction. How best to manage in this respect differs from case to case. It is in any event advisable for the supplier to identify crucial points in the cooperation process in advance and to initiate communication measures to prepare the customer for them.

The successful implementation of a solution into the customer's value chain strongly depends on cooperation between the employees of both the customer and supplier. It thus seems only natural for the customer to provide the supplier with required information and thus optimize cooperation, even though empirical evidence suggests that the latter tends to be very difficult: communication hurdles exist, important pieces of information are not delivered in time, responsibilities are left unclear, and budgets are exceeded. Should the customer company have brought such a situation about, its employees will either be unaware of the steps to be taken or refuse to cooperate. Ultimately, the source of the problem may well lie in communication by the supplier; for instance insufficient preparation of either the project or the general foundations for trust. The fact that providing confidential information makes a customer feel vulnerable here needs to be taken into account. This is especially the case when a supplier also offers services to a customer's competitors. The supplier has to anticipate and resolve such problems – the key to this again lying in building a foundation of trust.

In sum, the major goal of communication in the post-sales phase is to successfully guide cooperation between customer and supplier and thus optimize the solution's integration. For this reason, it is necessary to obtain detailed information about the customer's value chain, to agree on the changes that are to be made to successfully manage implementation of the solution, and to maintain an image of reliability.

Communication during the Up-Sales Phase

As already indicated, the complexities involved in most provided solutions means that customer-supplier cooperation will be required even after the post-sales phase. In order to optimize their transaction costs, customers have a great interest in such cooperation. But the supplier benefits even more from this continued cooperation. Practical experience tends to confirm the principle that it is less cost intensive for a supplier to do business with an existing customer than to acquire a new one. Therefore, maintaining customer loyalty is not only an objective of communication policy but also an economic imperative for a supplier in the up-sales phase.

Especially when maintenance and repair requirements assure further contact, suppliers will have a good opportunity to familiarize themselves with their customers. This familiarity then provides an excellent basis for identifying potential improvements in the customer's value chain and thus developing new needs for the customer. To this extent communication in the after-sales phase should always be directed towards further sales.

The newly identified sales potential is at the supplier's disposal, provided that the customer is satisfied with the initial service. Consequently, the supplier should regularly inquire about the customer's satisfaction with the product and, in general, orient the communication towards influencing the satisfaction level. After the purchased equipment has been installed, the customer will again be fully focused on day to day business and will thus hardly pay any attention to the transaction's – hopefully – positive effects. Should the customer be satisfied with agreements held, promises kept, and goals achieved, the supplier should aim at a continued awareness, as it were, of the satisfaction.

Finally, it should be noted that communication in the up-sales phase should not be directed solely towards future purchasing assignments. In order to enhance the relationship with the customer, personal relations with the customer's employees that go beyond standard business relations should also be established.

Fig. 1 summarizes the communication objectives of a supplier in B2B markets as well as the sub-targets in the four phases.

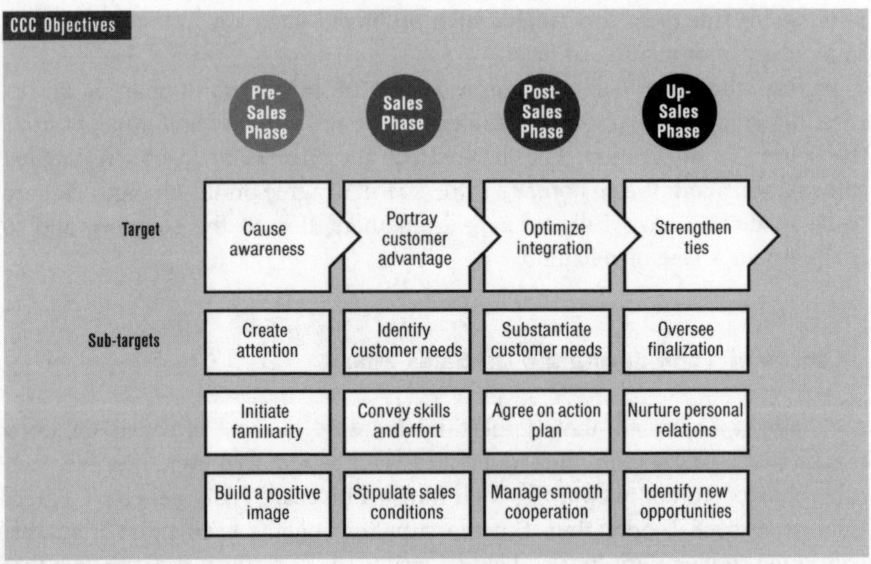

Fig. 1: CCC Objectives

The CCC Phase model

Theoretically the sales process is categorized by specific phases, yet practically a clear differentiation between the four phases is not always apparent. The transition from the sale to the post-sale phase can take place in various ways and in a variety of steps, and not necessarily by contract-signing alone. To distinguish precisely between post-sales and up-sales is equally impossible: when, for instance, has maintenance work on aircraft turbine paddles been completed? When the paddles have been checked and replaced? After the motor has been tested? When the motor has been replaced in the aircraft and the aircraft is once more active? In a similar manner the phase objectives may themselves not be clearly separated, since the after-sales phase may influence customer behavior as much as product advantages seen in the pre-sales and sales phases; but each phase is necessarily tied to the next, since each after the first can only be realized once the earlier phase has been completed.

The progressive structure of the CCC model shows that the objectives of already-concluded phases continue to play an important role. Therefore, suppliers should not consider their communication measures as isolated activities, but rather perceive them as a coherent whole covering the four phases. All individual measures should complete each other and, seen from the customer's perspective, be cohesive. This is above all the case for formal aspects of the communication process such as a consistent use of the company's logo in non-personal communication.

Even more important is the fact that the information conveyed in all four phases should contain neither inconsistencies nor contradictions. On a theoretical level this observation may seem patronizing. Practical experience has

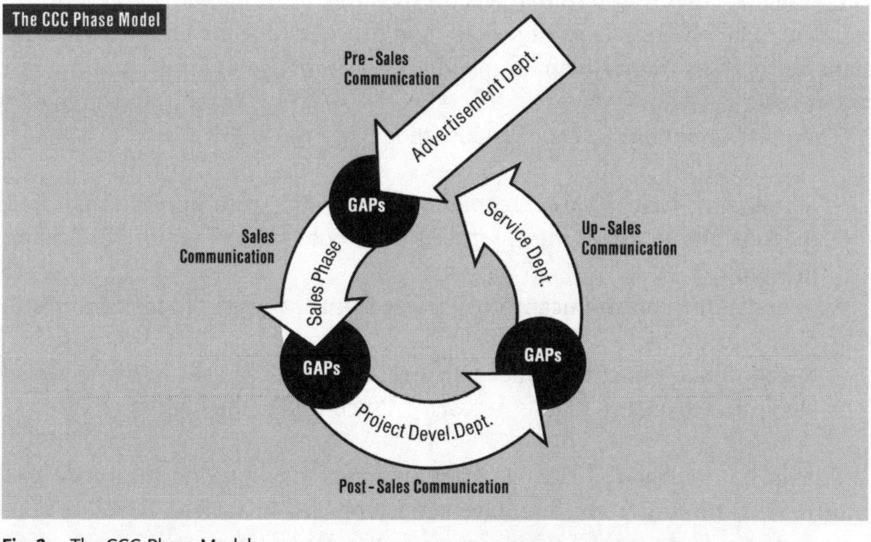

Fig. 2: The CCC Phase Model

shown, however, that it is by no means self-evident. For example, a service employee of a machine-building company might inform a customer that the need to repair a certain machine is the consequence of ordinary wear and tear. The customer might then react with surprise and unhappiness because a sales representative told him before the sale that in the first two years the machine would be free from any repair caused by standard usage. Such a communication mishap is by no means exceptional; the mishaps are not a result of coincidence but mostly based on systematic mistakes in the supplier's communication strategy. For instance all large industrial corporations divide the customer communication in the four phases among various people with various responsibilities. The advertisement department is mostly concerned with communication in the pre-sales phase, whereas the sales department deals with the sales phase. The cooperation with the customer in the post-sales phase is handled by the responsible project and development engineers, while the service department (maintenance and repair) or the sales department (personal contact) arrange communication in the up-sales phase.[102] Often gaps or even reluctance among the departments involved hamper coordinated communication with the customer.

These organizational difficulties are often exacerbated by the prevalent understanding of marketing communication. As shown above, in most companies as well as in relevant textbooks, communication only refers to activities in the pre-sales phase, occasionally together with a few activities in the sales phase. This clearly too-narrow perspective is bound to result in communicational mishaps with regards to the customer. To continue the above example: service employees only feel responsible for repairing customers' machines, not for realizing communicative objectives – it is very likely they do not even know them.

Given this situation, we clearly need to develop a new approach to communication management. In B2B markets this means extending marketing communication to all phases of a sales process and providing for the necessary coordination processes (rather than simply dividing up budgets, funds, and human resources).

The most important facets of a CCC model are the following:

- to overcome a narrow understanding of marketing communication in B2B markets and to direct the communicative policy to the entire purchasing process;
- to derive the communicative objectives from the sales phases of the supplier's company;
- to coordinate the communication measures in all phases into a cohesive whole and to ensure that its deployment will lead to the highest efficiency.

It should be emphasized that such an approach goes beyond the models of integrated communication that have been proposed in various business contexts over the past years.[103] With regards to the American context, the work of

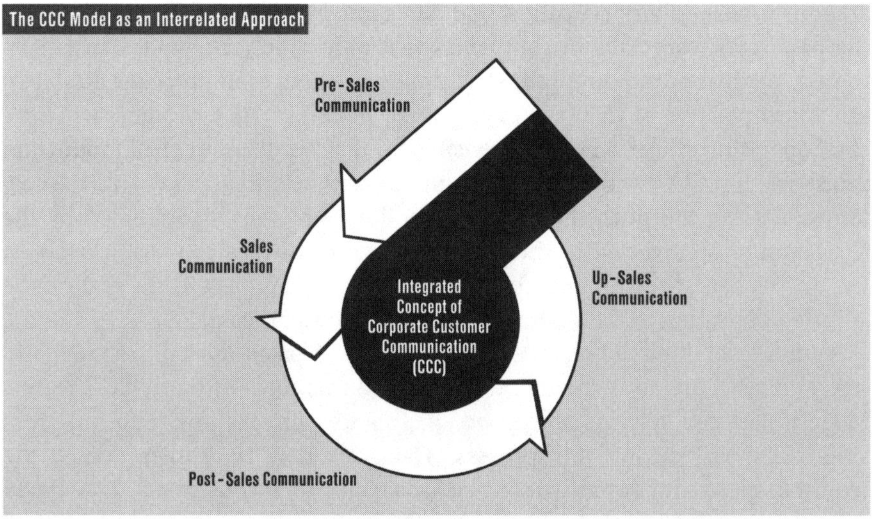

Fig. 3: The CCC Model as an Interrelated Approach

Anderson and Narus here comes to mind: their model of integrated marketing communication, postulating a "seamless integration" of all communication measures, clearly remains within the familiar boundaries of instruments such as advertisement, sales promotion, public relations, and so forth.[104] The same can be said for the work of Dwyer and Tanner, whose model of the same form of communication stresses the importance of two-way communication in B2B markets but then proposes implementation methods that are again instrument-based.[105]In Europe Bruhn has developed a similar approach,[106] adding the dimension of follow-up transactions and customer relations to his discussion of communication policy but insufficiently considering communication processes important in B2B markets, such as reactions to public advertisements, contract negotiations, coordination of customer cooperation, and support of after-sales services. These customer-directed communication processes represent the core of the CCC model. All such processes need to be taken into account and coordinated, especially when they have to be organized across both individual departments and – in light of the international activities of many companies – national borders.

Challenges and Solutions for the CCC Model

In business practice, this model points to the challenge involved in managing a highly complex process. The structure of a company should be adjusted accordingly. Möller and Rajala thus recommend that high-tech companies develop so-called marketing centers to coordinate the intensive internal communication process.[107] Often, this coordination is the responsibility of a key account manager: an occupation that has gained strongly in importance over the past years, especially in companies that only supply a few very large customers, such as automotive industry suppliers. Sheth even proposes establishing a new position of CCO (chief customer officer).[108] As a counterpart to the chief operating officer, who is responsible for the coordination of all production processes, the CCO would direct all communicative processes towards the customer. During the procurement process, the employees, coordinated by the CCO, would be involved in the communication activities to varying degrees. For this reason, it seems advisable to develop flexible project teams that can be adjusted according to the requirements of the sales process.

A high level of clarity regarding both measures taken during the sales process and reactions to it should be guaranteed members of these CCC teams. Modern database systems offer good chances to overcome any technical problems occurring during this process. Data can now be transmitted at the required speed and saved without redundancies or inconsistencies. A broad spectrum of solutions addressing these CCC expectations has been developed under the rubric "CRM" (customer relationship management). The research community has also increasingly attributed importance to CRM; it is presently perceived as a key component in successful marketing.[109] Nevertheless, for many companies that now have CRM systems, the investment has not fulfilled the high expectations. Aside from conceptual and technical problems, the main reason for this situation is simple: a lack of acceptance by the employees.

As suggested, most employees will not perceive their job as part of a larger communicational structure. Substantial communication barriers are especially present among those employed in technical divisions. The barriers become even more formidable when the transparent system ordered from above is regarded as threatening. For example, the input of information into a CRM system not only means providing data considered valuable, but also openly displaying the work that has been done (and not necessarily flawlessly).

One solution to such challenges may be introducing what has been called a hypertext organization. Traditional organizational structures in a company follow a functional orientation, with business units, departments, and positions being created in mind of specialization into different company functions. This principle is usually apparent within the marketing realm, with different organizational entities specializing in different marketing sub-functions. It is likewise apparent within the sales realm, with almost every industrial company having a separate sales department. Media communication, service, and customer care are other typical fields of specialization and labor division within marketing. The major advantage of this functional approach lies in its enduring capacity to provide a foundation for organizational efficiency. Specialization leads to cost reduction and output improvement. But at the same time, the approach is marked by an incapacity to secure consistency for activities crossing an organization's functional borders: in particular, to allow communication in the sense of our CCC model.

Consequently, in the case of small or medium-sized companies where range and scope of business activities are restricted, a rather radical shift from a functional to an activity-based organizational structure might well be called for. The capacities of individuals or small units, though limited, might be sufficient to develop enough expertise for cross-functional activities and compensate for reduced degrees of functional specialization. In larger organizations, this is not the case, as activities are far more complex and numerous. Even standard business activities require that employees cooperate with others and divide the total amount of labor stemming from business processes. Introduced by the Japanese author I. Nonaka and later developed by others,[110] hypertext organization as a specific mode of internal organization here strives to combine the advantages of functionalist and activity-based organizational structures while avoiding the problems of both. The basic idea is illustrated in fig. 4.

As a permanent basis for a business organization, hypertext organization continues to rely on functional structures, division of labor, and specialization. But as a second layer, the company establishes a set of networks including people from different functions and different hierarchical levels. We need to note here that the designation *hypertext organization* was chosen in analogy to the way entities are connected in the hypertext mode of Internet documents: each network is in charge of a specific business activity requiring cross-functional coordination and bringing together expertise from each of the functions. Whether the network's working mode is formalized or follows informal guidelines depends on the specific company and business situation. Different networks may exist and individuals may be members of more than one of the networks. Marketing communication is probably predestined to be a primary field

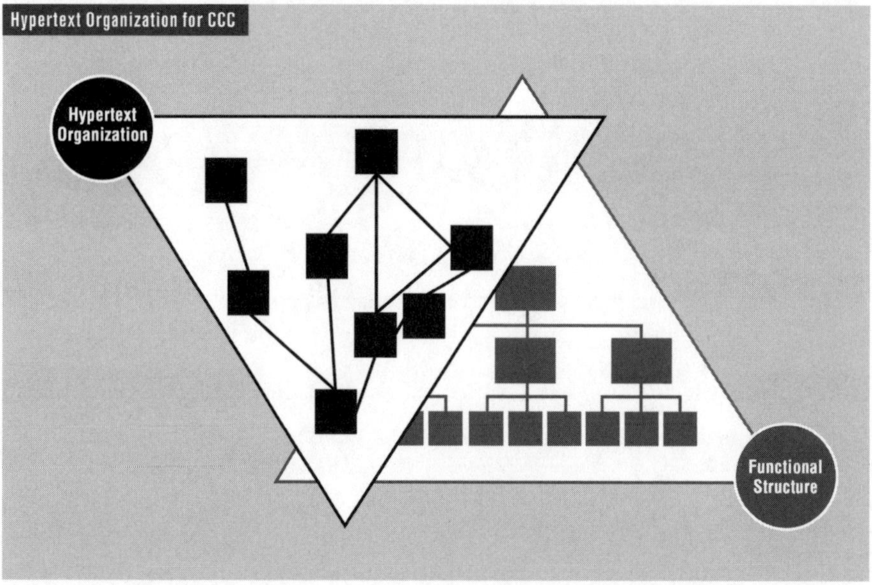

Fig. 4: Hypertext Organization for CCC

of application for this kind of network. Within marketing, work will continue to be organized along various functional lines – media communication, sales, service, customer care, and so forth. But for major projects, major customers, or very important market segments, a hypertext network will be available to coordinate activity consistency through the different phases of corporate customer communication, in other words through the pre-sales, sales, post-sales, and upsales phases. For the sake of both managing the network and facilitating its impact on day-to-day work, it seems advisable to implement a corresponding IT structure with a data base that collects all the relevant information and makes it available to all network members.

It is crucial to here keep in mind that even if employees are willing to adjust, the demands modern technology imposes on them may overstrain their abilities. Certainly, Ingram et al. have a point in stating that in the future we need intelligent sales managers who are skilled in hearing "the voice of the customer."[III] Following the approach proposed in this article, we might conclude that not only sales managers should meet this description, but rather each employee responsible for customer communication. However, in reality we are constrained by the fact that most employees simply do not possess the required technical or – and this is most important – social skills. Therefore, an implementation of the suggested approach to marketing communication should always be accompanied by development of appropriate skills and incentives for those who have to convert it into practice.

8
Leading the Sales Force

Robert E. Spekman
Benjamin Moor

Introduction

Critical to any attempt to bring new technology to the market is the ability of the sales force to work with both existing and new customers, educating them regarding the advantages of the new technology. In addition, the sales force, in conjunction with marketing, must develop value-added solutions that solve customers' problems and help them become more competitive in their different markets. But not all sales forces are equipped to understand customer problems; nor are all able to work effectively with customers, furnishing them with solutions recognized as providing superior value. In many instances, new technology is sold on the basis of price, and any premium is bargained away during the negotiation between buyer and seller. This inability to capture the value created for the marketplace raises the following questions: What role does the sales force play in bringing new technology to market, and what impact does effective sales leadership have in the sales process? That is, given an average sales force, what tools and techniques can sales management employ to improve its performance?

We believe that the role of the sales force is in transition and that the nature of sales management is itself in the midst of profound change. As it has been described in textbooks,[112] selling tends to imply an effort to influence the actions of another person to buy what you sell. For instance, it is said that a good salesperson can sell snow to an Eskimo. The observation's ironic nature notwithstanding, it is clear that in a competitive world such a sale would jeopardize the seller's longer-term relationship with the customer. In general, as sales forces do less selling and more problem solving, engage in less talking and more listening, and work to build longer-term relationships instead of engaging in transactional deals, the sales leadership can be expected to face greater challenges in its efforts to ensure positive outcomes.

In this chapter, we will first identify key trends that are both changing the way companies sell and making the role of sales leadership a tumultuous one. In light of these trends, we will then discuss objectives for developing an adaptive and relationship-driven sales force. Finally, we will present the sales leadership processes necessary to create a sustainable and effective sales force.

Bringing Technology to Market. Olaf Plötner and Robert E. Spekman

ISBN 978-3-527-50270-7

Key Trends

The globalization of markets, rising levels of competition, increasingly demanding customers, and the ready availability of information are forcing companies to be inventive in the ways they maximize value creation. Traditional product companies are being forced to sell not only the bundle of attributes that constitute their products, but also the bundle of relationships that constitute the partnership experience. When a selling organization converts a buyer-seller arrangement into a collaborative relationship, it has created an intangible competitive advantage. Efficiently and effectively crafting these relationships in light of a variety of trends will be the greatest challenge facing the leaders of sales organizations.

The growing complexity of the selling environment is putting a tremendous strain on salespeople. Customer expectations continue to ratchet upward. Salespeople are required to have a solid understanding of new accounts before initiating contact, and they must respond to requests in an increasingly rapid fashion. Other major complexity issues include the acceleration of changes in new product development and technology, buying center interactions involving more influencers, and greater diversity among customers as sales organizations become more global in scope.[113] A relationship-driven sales department places much of the pressures of this change upon the shoulders of sales representatives and account managers. As customer demands evolve, front-line salespeople are required to constantly obtain and assimilate new information to remain effective. Cognitive demands and greater uncertainty create a more stressful role, but also an opportunity for creativity and thoughtful leadership at the sales-representative level.

There are no products, only solutions is the new mantra. For instance, the industrial valves sold to customers might be state-of-the-art: tolerant of high pressures, resistant to corrosive materials, and compatible with industry-standard flow controls. But even a highly specialized feature set is often not enough to make a sale. A customer might want favorable payment and delivery terms, low defect rates, high service levels, a broad variety from which to choose, two-hour service-response time, spare parts delivery within an hour, backward compatibility, or any of hundreds of other possible service elements. Recently, companies have sought to identify and offer those value-added services with the potential to maximize the competitiveness of their product offerings. The more an organization can sell the partnership experience with all of these attributes, the better positioned it will be to avoid the margin-eroding menace of commoditization.

The enterprise server market is a good example of a competitive market that has recognized the need to bundle value-added services together with their product offerings. IBM, HP, Sun, and Dell all seek to protect their lucrative relationships with large data centers and thus have devised a menu of services to suit the needs of these valued customers. Fig. 1 illustrates our argument

An Example of Value-Added Services as Part of a Solutions-Based Sales Effort

Planning Services	· Hardware capacity planning · Roadmap
Break-Fix Services	· On-premises technician · On-site spare parts · 1-day, 4-hour, 2-hour parts delivery · Proactive repair with automated "call home" features
Support Services	· 24x7x365 Telephone support · Phone support guaranteed under one hour · Dedicated account manager for telephone support · Access to repository of technical articles · Software patches and updates · Software patch installation · Support for critical software (OS, Email, ERP, etc.)
Service-Level Agreements	· 99.999% uptime guarantees ("Five 9's")

Fig. 1: An Example of Value-Added Services as Part of a Solutions-Based Sales Effort

that the server market has become far more complex than selling the right item.

Sales organizations are seeking to be more relationship oriented, adaptive, and agile. The new sales management paradigm requires sales leaders to strive to maintain a high level of efficiency and effectiveness while maximizing their organization's adaptability and agility. Efficiency and effectiveness are measures of a sales force's ability to convert dollars into sales calls and sales calls into revenue. These goals typically are addressed with the traditional levers: performance-based compensation, territory alignment, adjustments to the product mix, and control mechanisms. But longer selling cycles, greater reliance on team selling, and less ability to measure a direct link between sales effort and a sale have called into question the metrics used to measure performance and reward appropriate sales behavior.

Changing Roles and Different Demands

It is clear that these increasingly complex dynamics have placed a burden on the sales force. It must coordinate efforts across functions and disciplines within its firm and must integrate those efforts with corresponding functions and units in the buying organization. To complicate matters, the sales force has little authority for organizing this level of integration and coordination. Given the speed of change and global competitive pressures, time is indeed a rare commodity. The need for the sales organization to demonstrate adaptability and agility compounds an already difficult challenge for sales management and would seem to contribute to role stress and role overload.[114] Adaptability here refers to a salesperson's ability to change strategies during a sales call as new

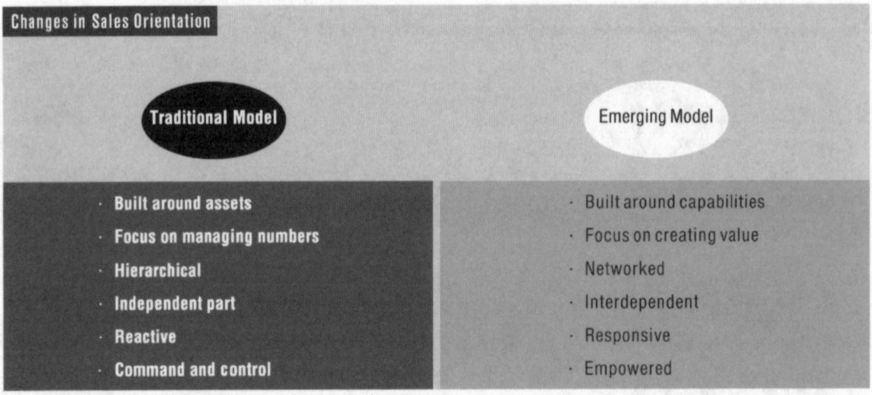

Fig. 2: Changes in Sales Orientation

information becomes available. Agility refers to the application of knowledge of company resources and capabilities to creatively and expeditiously offer customers new solutions.[115]

A growing body of research has shown that the higher the degree of relationship between a salesperson and a customer, the greater the salesperson's chances of convincing the customer to either keep its existing products or purchase new products from the same company. A successful strategic supplier-buyer relationship can provide partners with access to a variety of resources – financial, physical, informational, and relational, among many others.[116] Companies are seeking means to develop and deploy the skills required to leverage their sales forces as a source of competitive advantage.

Fig. 2 illustrates the types of changes that are needed as firms begin their transition to becoming more responsive, customer-centric sales organizations. Notably, emphasis is here placed on working as part of a network of cooperating companies sharing common goals and objectives, acknowledging mutual dependence, and as a consequence exchanging information in real time. Sales management must itself reflect that transition and work hard to encourage corresponding sales behaviors. The pressures facing the sales force can easily be imagined: the boundary-role individuals involved must manage across units and companies, form teams whose members typically see themselves as individual contributors, and develop behaviors that support working with customers as trusted advisors and honest brokers. To support these new expectations, sales management must hire and train a different type of salesperson, motivate and reward different types of behaviors, and focus on behavior-based systems encouraging processes and procedures that achieve a desired end.[117]

Fig. 2 allows us to understand the change in approach required to address this different perspective. At the same time, a number of activities must be performed if the sales process is to be effective. Dubinsky has divided the sales process into units of essential work he refers to as the "seven steps of selling": prospecting, pre-approach, approach, presentation, overcoming objections, close, and follow-up.[118] But selling has evolved in a way meaning that most companies no longer follow such a sequential pattern.[119] While essential to the completion of the sales process, not all activities are within the salesperson's work scope, since a number of the activities can (or should) be performed by others. This shift in responsibility is intended to free the sales force from more mundane tasks and focus attention on those activities more likely to result in a sale. Prospecting, for example, is now frequently conducted by telesales, the marketing department, or even the Internet in order to reserve the sales force for high value-adding efforts where the probability of a sale are higher. Put simply, the objective is to maximize sales force effectiveness and efficiency. Often one finds some form of tiered structure where there are people and processes devoted to lead-generation and moving prospective buyers to the point where they are ready to buy. These structures reflect an understanding of sales productivity where the goal is to protect the sales force from those lower-priority tasks that are still essential to the sales process. Fig. 3 provides a sample division of essential work among different selling channels.

Several insights emerge from fig. 3. First, there is no single best sales channel for implementing a successful sales strategy, as one must compare effectiveness and cost. Also, it would appear that information technology plays a critical role in the process, since a number of channels complement the sales effort and can be implemented for a fraction of the cost of a salesperson. The Internet has been shown to be a perfect vehicle for qualifying leads and assessing the degree to which an inquiry can be converted to a sale. Sales automation

Different Roles Played by Different Sales Channels

	Missionary Sales	Telesales	Outside Sales Key acct mgr	Tele-prospecting	Internet
Lead Generation		X		X	X
Quality		X		X	X
Account Planning	X		X		
Opportunity Development	X		X		
Tailor Value Proposition	X		X		
Provide Solution	X		X		
Negotiate & Close	X		X		
Terms & Conditions			X		
Implement			X		

Fig. 3: Different Roles Played by Different Sales Channels[120]
Source: Zoltners, et al: Sales Force Design

software has provided an organizing framework for both salespeople and managers to track progress of a sale effort, assess whether goals are being met, and highlight areas in which coaching and remedial work are appropriate.

Information technology is becoming much more central and prevalent among sales organizations. Sales force automation (SFA) systems, customer relationship management (CRM) systems, and other technologies are providing sales organizations with valuable functionality–but are also bringing new challenges. Selling productivity improvements due to increased technology includes substituting capital for labor, changes in the sales job, reallocation of tasks within the buyer-seller interaction, improvements in team selling through stronger communication, and improvements in the organization of the selling function itself.[121] The technology has absorbed routine and easily automated tasks to free up time for salespeople to concentrate on more interaction-intensive tasks like problem solving and account planning.

Salespeople and sales management tend to view the gains from SFA differently.[122] Sales managers have a tendency to perceive greater productivity gains and be more satisfied with SFA tools. Salespeople tend to view SFA as a means of micromanaging and are unsatisfied with the amount of information provided by the system. At the same time, the functionality of SFA or CRM systems varies dramatically across vendors; these systems' effects on redefining sales roles and increasing communications will likewise vary. Knowledge-management-based SFA systems are capable of dynamically pushing relevant information out to the distributed members of a sales force. SFA solutions also offer sales tracking and relationship management. These systems have been shown to improve the performance of individual sales representatives.[123]

Given the mixed degree of acceptance of these tools, it should not be surprising to learn that the failure rates for CRM installations are high, ranging between 60 percent and 80 percent.[124] Kale has cited the following top seven reasons for an unsatisfactory CRM outcome: viewing the CRM initiative as a technology initiative, lack of customer-centric vision, insufficient appreciation of customer lifetime value, inadequate support from top management, underestimating the importance of change management, failing to re-engineer business processes, and underestimating the difficulties involved in data mining and data integration. A 2002 survey conducted by DMR Consulting (now Fujitsu Consulting) revealed that nearly two-thirds of respondents claim that their company is no closer to being customer-centric after a CRM installation than before.[125]

Despite the normative appeal of sales automation tools, in general they have over-promised and under-delivered. While it is intuitive to assume increased sales force effectiveness and efficiency, the data shows that advances in technology result in disconfirmed expectations.[126] The issues germane to understanding the failure of such programs seem to converge on those factors affecting buy-in and utilization by the sales force. Factors affecting adoption and utilization of the technology range from cost, training, and sales force involvement in the decision to buy technology, to personal factors and a lack of sales management support. Bush et al. have suggested that adoption of sales automation tools is driven by perceived usefulness of a tool and ease of use. Interestingly, since the tool facilitates cross-functional integration and the importance of enterprise focus on the customer, buy-in and commitment to the tool should not be limited to the sales function only, rather requiring organization-wide involvement. To be sure, this technology represents a significant transition as the firm becomes more customer-centric, although it must be acknowledged that any sales automation tool is only an enabler: the key to success is the ability to adapt organizational processes and structures so that they are better aligned with the new strategic goals of the firm.

Accompanying the advent of new technologies is a realization that not all customers are created equally and that it makes sense to have structured sales teams around individual customers or sets of customers. In certain circumstances these customers are critical because of their revenue impact (global/national account management) or for other, strategic reasons (strategic account management). Members of key account teams typically represent different functional areas, ensuring the team is providing comprehensive customer solutions and that internal functional goals are met. Research suggests that key account management yields numerous benefits, including greater effectiveness, faster responsiveness, smoother communication, and ultimately stronger relationships.[127] The goal of a strategic account team or core selling team is to bring buyers into the selling process, solicit their feedback, and eventually meet their most critical needs.

To complicate matters further, it can be expected that a number of these account teams will be assembled to address the demands of global customers. It would not be outlandish for a global firm to ask its supplier to roll up all of its invoices on a worldwide basis, issue one monthly global account invoice, and have the customer select the payment currency. Those demands seem aligned with other considerations tied to the physical flow of goods; and with problems associated with linguistic and cultural differences, separation of time and place, and the challenge of presenting global customers with a unified appreciation of the seller's value proposition and close relationship.

What Drives the Process of Becoming a Solutions Seller?

The first requirement for effective solution selling is the development of a *culture* driven by customer needs. Adopting customer orientation requires a redefinition of sales roles to explicitly focus on a broad understanding of the customer's business, economics, and specific circumstances. At the core of the process is the ability to communicate value and to present a unique solution meeting present and future needs. One way to do this is to bring together other features, partners, and services. Such bundling creates a unique market position and can create barriers. Sellers can develop programs that meet customers' buying patterns and needs (warrantee programs or consulting services), make one's firm easier to do business with, and allow the customer to be served at lower costs.

Nevertheless many firms are dissatisfied with SF effectiveness and efficiency. In fact, a Deloitte study has shown that 60 percent of companies question their sales forces' productivity. To make the value model work, the definition of value must be derived from the customers and not internally. Customer value must be linked to the firm's core skills; if not, there will be a major disconnect between what is competence and what is valued. Hard work by the firm to incorporate the customer's voice is critical to the process.

Concomitant with the need to provide value is a recalibration of performance and a shift away from a focus on top-line growth to a focus on the lifetime value of a customer. Sales management thus must evaluate every customer as a potentially valuable annuity with calculable present value. Sales organizations must adopt a long-term view in order to gain customers with large future payoffs; they must not waste their sales effort on potential customers who will be a net drain on resources. There is a potential tension between evaluating the worth of a customer and providing key services. For instance, the lifetime value calculation for a buyer who asks for MRO inventory to be held on premises, with invoicing based on usage, would be reduced by the holding costs of the consigned inventory.

Achieving the level of relationship selling required to affect a high positive net present value partly involves a great degree of interdepartmental collaboration, especially between the sales and marketing departments. However, in a number of firms these departments have traditionally been at odds, tend to distrust each other, and often are measured by different metrics. Companies that seek to maximize customer value must take an enterprise view, seeing every point of customer contact as an element in the product and service mix. Full value cannot be conveyed to the customer if functional units do not cooperate.

Companies can take three structural approaches to increasing integration between sales and marketing: decentralization, cross-functional teams, or cross-functional integrator roles.[128] Frequently, the sales and marketing departments are completely subsumed to their respective vice presidents and only come together at the top of the hierarchy. To address these issues organizations can create a single head of sales and marketing, or else implement lower-level integrator roles to connect salespeople and marketers. Given appropriate incentives, composition, and tasks, cross-functional teams can be another means of increasing sales-marketing integration.

It is in any case clear that these linkages are not natural and are often subject to conflict and tension. One solution to this is the use of implementation teams that can add differential value by proactively solving compatibility issues or suggesting additional ways to leverage the customer's investment. Here, the account team bears responsibility for monitoring these customer touch-points and providing technology road-mapping and advisory services. This is a point where salespeople can build long-term customer relationships as trusted brokers.

Merely providing information does not provide value; in that case the salesperson becomes something like a talking brochure. As indicated, value is added when the sales team becomes an advocate of the customer or provides insight into the needs of the customer's customer. Problem solving is generally considered very important to the customer's value equation. Notably, this customer expectation has placed a burden on the sales force and on sales management such that now the salesperson must be seen as a resource for senior management and as a contributor at the C-level (COO, CMO, CIO). If we are concerned about transforming the salesperson into an effective solutions provider, then establishing a credible presence at the C-level becomes a serious challenge for sales management.

Effective Sales Management

What impact does sales management have on the role played by the sales force and what levers can sales leadership pull to enact change and gain a desired outcome? Given the evolving focus on relationship marketing and consultative selling, these questions become quite salient. The questions can be addressed in the context of how sales management leads the sales force. Sales leadership within a relationship/consultation paradigm is rather complex, since one must account for traditional problems: territory alignment, top-line growth, soothing egos and personalities and reigning them in; all this together with developing a business model that encourages bottom-line results, or put otherwise, building a customer-centric organization in which an enterprise-wide perspective supplants silo-like thinking. Sales managers become conduits for information, critics and guides, trainers and coaches. To be sure, they still hire, train, deploy, reward, and motivate salespeople, but in the context of changes in the business environment that have altered the nature of the relationship between salespeople and their company, more attention must be given to how to improve the alignment between the firm and the sales force. Hence there is a new emphasis on raising the level of mutual goals and objectives.[129]

Cron et al. have discussed the range of knowledge, skills, and abilities (KSAs) that are essential to achieving a high level of sales effectiveness. Baseline skills are task related and highlight those minimal (albeit critical) areas such as product/market knowledge, selling and closing skills, territory/time management and the like. Growth skills include such considerations as an ability to learn, adaptive behaviors, problem-solving capability, and goal-setting skills. Finally, meta-skills emphasize self-development, self-assessment, and self-regulation. These skills enable the sales force to grow and develop a stronger and more appropriate set of task-related and growth skills.

Specialists in this area of research distinguish between transactional leadership and transformational leadership. Transactional behavior hinges upon distributing rewards and punishments linked to salesperson performance.[130] Transformational leadership involves fundamentally changing the values, goals, and aspirations of employees, so that they perform their work because it is consistent with their values and not primarily from an expectation that they will be rewarded for their efforts.[131] Transformational sales leaders provide their sales forces with a vision, appropriate role models, individualized support, intellectual stimulation, a group goal orientation, and high performance expectations. These attributes are presented in fig. 4. It should be noted that the actions shown on the right re-enforce the KSAs discussed previously. The goal is to produce a self-directed, highly motivated, high-performance sales team.

An Illustration of Transformational Leadership

Transformational Leadership Attributes	Transformational Leadership Actions
Articulate a vision	· Consistently live and breathe the mission · Repeat and reinforce the mission · Be relentless in defending the vision internally and externally · Simplify and clarify the salesperson's goal · Be receptive to new ideas
Provide an appropriate model	· Identify and exemplify past successes done the right way · Provide sales tools appropriate to maintaining relationships · Assemble teams with skill-sets to match customer needs
Foster acceptance of group goals	· Stress the importance of teamwork · Take team roles · Provide team incentives
Offer individualized support	· Provide assistance in approaching customers whose needs are difficult to identify and address · Mentorship
Encourage intellectual stimulation	· Push critical information to the sales force · Monitor and pass on high-level knowledge of industry and · economic changes

Fig. 4: An Illustration of Transformational Leadership[132]
Source: McKenzie et al: Transformational and Transactional Leadership

The transformational model is inspiring, empowering, and motivational.[133] Empowerment is an important element in developing a responsive and customer-oriented sales force. Empowered salespeople feel capable of adapting their strategies to suit the specific conditions provided by each selling opportunity. Leaders create an empowering environment by allowing salespeople to participate in more decisions, thus reducing bureaucracy, and to be flexible as circumstances warrant.[134] But despite the lofty goals and aspirations of transformational leaders, salespeople still respond to financial incentives, and compensation plans are still a key element in driving desired behavior, producing superior results, and generally reflecting the firm's priorities. Positive reinforcement is an important element in motivating a sales force. The most common such reinforcement is cash (it is used 58 percent of the time).[135] But it is increasingly acknowledged that non-financial rewards are themselves a powerful influence on a salesperson's organizational commitment and goal acceptance.[136]

In 2003, 42 percent of companies changed their compensation plans, with almost half of these changes being modifications to performance measures.[137] This indicates that firms are re-evaluating the metrics that are most closely associated with sales force success. Traditionally, companies have offered rewards based on sales targets and other company results, in other words as a measure of output. Certainly, we have witnessed quotas, commission structures, and bonus plans that have been developed to achieve profit and sales goals within a line of products or across key segments or customers. However, when selling cycles are long and the linkage between a sale and sales effort is diffuse and difficult to attribute to a single person or a series of sales calls, output- and outcome-based models are less useful.[138]

In recent years, the emphasis on more collaborative, relationship-based selling strategies has called into question the more traditional outcome-based models and has sparked a great deal of focus on the most effective approaches to coordinating and managing the selling process.[139] To align salespeople's goals with a strategy of relationship building, a sales leader should provide incentives reinforcing this objective. Relationship selling requires a new set of performance metrics to reward salespeople for finding the *right* customers (with the greatest lifetime value), building share of wallet through up-selling and cross-selling, retaining key customers, and improving customer satisfaction. Fig. 5 summarizes the difference between traditional and new selling environments with respect to setting goals and rewarding a sales force. This figure is based on two research projects that examined the differences between behavior-based and outcome-based control[140] and formal and informal controls.[141] Although the language used in the two studies in question is different, the authors base their theses on very similar principles. Behavior-based control is characterized by an emphasis on the sales person's actions and input into the sales process. Informal control is based on social, cultural, and worker-initiated mechanisms serving to modify behavior in a socially positive direction within a company framework. In both instances, the belief is that certain behaviors will lead to certain outcomes. In addition, the percentage of fixed versus variable compensation will vary depending on the emphasis given to behavioral or informal control versus outcome or formal control. Specifically, variable compensation is tied to more outcome or formal control, while having a fixed salary is associated more with behavioral or informal mechanisms.

With greater emphasis given to behavioral and informal controls, it follows that both selection and training become more important for the sales leader. With the expectation that certain sales behavior will lead to a desired outcome, sales management must not only hire people who fit a profile or possess cer-

Comparison between Traditional Selling and the New Selling Environment	Traditional Selling	New Selling
Metrics	· Individualistic · Transaction-specific · Product-focused outcomes/behaviors · Sales function-oriented	· Team-based · Customer relationship-oriented · Customer-focused /solution-based · Cross-functional team-oriented
Goals	· Sales person and outcome-based	· Life time value of the customer/cost to serve · Customer satisfaction
Controls	· Individual sales outcome-based · Formal controls emphasized	· Cross-functional team-oriented · Informal controls emphasized
Compensation	· Sales person focused · Output driven · Quota based on shorter term goals	· Team or cross-functionally based · Tends to be behaviorally driven · Longer term goals focused

Fig. 5: Comparison between Traditional Selling and the New Selling Environment[142]
Source: Churchill, et al: The Determinants of Sales Force Performance

tain characteristics; it must also model this behavior, practice developing it further, and modify incentives to encourage this development.

There have been a number of studies exploring predictors of sales performance; one of the most influential is that of Churchill, Ford, Hartley, and Walker.[143] The range of predictors extends from skill and knowledge assessment to cognitive ability and personality inventories to personal histories and targeted sales assessments. It is safe to say that the results are inconclusive at best. For instance, cognitive ability as measured by IQ has been shown to be a strong predictor of performance; however, other work has demonstrated that IQ might not be as valid a predictor as previously thought and that emotional intelligence and practical intelligence also play an important role.[144]

Given the globalization of business, the importance of cross-functional teams, and the need to build direct and indirect ties among company networks, there are a number of challenges facing sales management. Among these is hiring and training salespeople who are adept at working in diverse cultural settings and possess similar skills to those of global managers. Such individuals will above all need to have a global orientation, which implies an understanding of global competition, an appreciation of why an enterprise perspective is important, and skill at balancing the local and global picture with a concern for detail. Moreover, they should be equally effective as team members and individual contributors. There is thus an inherent tension in the sets of skills and attributes needed to perform effectively.

Nevertheless some of this tension can be mitigated since successful salespeople will likely have a strong learning orientation.[145] For example, one study has suggested that in their desire to improve, salespeople with learning goals will seek to master tasks associated with long-term performance.[146] Hence growth-related training might help them develop problem-solving approaches, value recognition skills, coping techniques, and customer research skills. Many training programs only address task-related or behavior-related capabilities such as product knowledge and selling or analytical skills. Without management reinforcement through coaching and incentives, 87 percent of the new skills gained from behaviorally oriented training are forgotten within one month.[147] Adaptability, a key attribute in effectively addressing customer needs, has been shown to be more effectively developed through growth-related than task-related skills.[148]

In light of these challenges and pressures, sales management must approach coaching as a high priority and must spend enough time in the field to work effectively with the sales force. To be sure, span of control becomes a bottleneck to effective coaching; however, the ability to impart knowledge and demonstrate significant traits such as adaptability, agility, responsiveness, and empathy are important. Notably, these characteristics are not teachable but are learnable. For example, sales management can arrange opportunities to learn through rotational assignments, working closely with a more seasoned colleague, or providing exposure to task forces, and so forth. The advantage of this process is that each assignment has a coaching component. But realizing the advantage depends on overcoming a basic difficulty – that of creating a culture in which coaching is actually valued and rewarded. In fact, if a sales culture does not value coaching as the single most important activity performed by a sales leader, or does not tie coaching metrics to management rewards, or again does not search for ways to increase the time devoted to coaching, then it is very likely that the sales force will not be world class.

A related possibility is using mentors to improve performance, increase morale, and decrease turnover. Mentoring is not new to organizational thinking, but it has not been widely researched in a selling context. A mentor is a relatively experienced person (either a peer or someone higher in the company hierarchy) who is committed to facilitating upward movement and furnishing support to the career of another employee.[149] In the present business context, it is difficult to remove peers from their selling roles and have them work with colleagues, since this could be seen as introducing more non-selling time into a salesperson's already busy schedule. In addition, firms with a transactional mentality will not appreciate the value of social learning.[150] Here again, we can observe that the impact of the prevalent business culture is quite profound.

Conclusion

In a B2B context, it is widely accepted that sales plays a key role in adopting new technology and in the selling firm's ability to bring innovation to the marketplace. As the speed of change quickens, as technology becomes more costly, and as customers become more demanding and more global in their reach, the sales function also changes. Generally, we have seen that a key to success is the ability to develop strong and enduring ties with key customers.[151] These ties are built on the information flow between buyer and seller, together with a recognized interdependence and degree of trust between them. But not all sales forces are prepared to make the transition this entails, nor are all sales managers prepared to lead the transition.

One approach along the necessary lines is to provide solutions to customers and to be seen as an agent of the buying firm in the home office. Buyers buy solutions and answers, they do not buy products. Solutions imply a value-adding capacity that ultimately brings the buyer a relative advantage and sets a common vision for both buyer and seller. It is here that sales management must see through a significant change in how business has been done traditionally in many organizations. In even the most enlightened and progressive companies, becoming solutions focused and developing a customer orientation is anything but a trivial challenge.

Zablah, Johnston, and Bellenger have developed a model for the effective adoption of technology and for the emergence of partner-like attitudes linking buyer and seller in a common vision. While intended for use in another context, the premise remains valid in our own, pointing to the importance of the integrating role played by sales and sales management within both the firm and larger network of which it is a part. Through its direct and indirect ties, sales is responsible for bringing customers technology that adds value, satisfies customer needs, and offers them a relative competitive advantage. Given the complexities of a global competitive arena, the sales task is complicated by the need to spread resources over key customers located all over the world. The ability to integrate functions internally in order to provide key products and services is as daunting a task as trying to deliver these value-adding capabilities globally.

In this discussion, we have tried to underscore many of the issues and concerns driving the sales process and resulting in a more collaborative relationship between buyer and seller. We have also outlined the challenges facing sales management as the sales role changes from communicating the value received by the customer to becoming part of the value creation process itself. These changes require a new set of metrics to reflect the importance of the new relationships. We have thus discussed the lifetime value of customers and the need to look beyond sales and margins: to calculate a return on investment reflecting the investment made in developing and nurturing the new relationships.

Coupled with the need for new metrics is an understanding that the solutions-focused or customer-centric sales model is built on a new set of behaviors and capabilities on the part of the salesperson. We have reviewed the challenges facing sales leaders as they offer incentives to, motivate, and coach the twenty-first-century sales force. To summarize our central argument: Developing a global mindset together with a consultative approach to selling involves a number of challenges, not least of which is the natural tension emerging because of apparently inconsistent demands and goals. Instead of constantly selling, the salesperson must listen, advise, and be the honest broker providing customers with the best solution. In some instances, the firm might not provide that solution. In short, change is the only constant; the effective sales leadership must acknowledge this change and have the courage to adapt. To this end, we offer several suggestions below for improving the effectiveness of the sales effort:

- Develop a focused sales strategy that entails directing one's effort at priority segments and customers. The effort must be aligned with the firm's strategic intent.
- Understand that organizational structure and process both enable and inhibit the best of plans. Develop plans that enable process and structure and have metrics that are aligned with and support the plan.
- Be aware of the need to motivate and encourage the sales effort.
- Be consistent in recruiting, training, motivating, and working with the sales force so that it focuses intently on customer-centric value-adding solutions. In addition, recognize that success hinges on these attitudes being seen as paramount when it comes to company culture, to be shared by all its employees.

As a final comment, we would suggest that the goal of developing an orientation towards fully engaged customer-centric solutions is basically clear cut. The devil is in the details. But despite the difficulties involved in seeing through the transformative process, the effort is highly worthwhile: those sales leaders who understand the process and are determined to stay the course are likely to succeed. We hope to have provided direction and tools to ease this journey.

9
Measuring Sales Force Performance

Martin Kupp

Looking at the various trends and developments that shape the modern sales management of industrial goods and services as discussed in this book, one common denominator emerges: the need for efficiency reigning in the procurement and operations departments for so many years has now reached the marketing and sales department. The sales force is the interface between the company and its customers; often, especially in B2B markets, it alone is responsible for generating sales. A company's performance is thus very much driven by the performance of the sales force, and an accurate measurement of sales force performance will help us understand and predict company performance.

It is commonly agreed that sales organizations are responsible for implementing business and marketing strategies that in turn have an impact on the effectiveness of the sales organization. It is thus essential that sales executives understand how and to what extent these strategies, in combination with management control and salesperson performance, influence the performance of their sales organization. In many companies, particularly in B2B markets, the sales force accounts for the largest portion of both marketing personnel and the marketing budget. Therefore, in working to gain efficiency, controlling the sales force and selling activities with the help of performance indicators is critically important to sales executives.

Modern incentive systems are another reason why such indicators are often so crucial. Salespeople and sales managers are usually the employees with the highest proportion of variable income within a company. But when incentives play an important role, one has to know what their basis should be: revenues, contribution margin, profit, value creation, or the number of new customers, key accounts, and so forth. At the same time, management needs some basis for facilitating decision making in order to improve salesperson performance and enhance sales-organization effectiveness. This is especially true when a start is made at combining single control elements into a holistic control system. For all these reasons, the design and implementation of effective sales force measurement systems is a basic element in increasing the performance of the sales staff and thus of the entire company.

Bringing Technology to Market. Olaf Plötner and Robert E. Spekman

ISBN 978-3-527-50270-7

Outcome Performance versus Behavior Performance

Guided by approaches such as agency theory and transaction cost analysis, behavior-based and outcome-based control systems were identified as polar managerial strategies in the late 1980s.[152] Until now, research results on the actual impact of both outcome- and behavior-based performance measurement systems have been highly ambiguous.[153] For instance, outcome-based systems have been found to increase performance,[154] to decrease it,[155] and to have no impact on it at all.[156] Similarly, some studies have found that behavior-based control systems improve salesperson performance,[157] others that they are unrelated to such performance.[158] In any event there have thus far been very few efforts to examine the basis for designing sales force control systems. The purpose of this discussion is thus to suggest a practical framework for choosing either an appropriate sales performance control system or individual performance indicators.

As stated, such indicators can be classified in terms of either outcome or behavior. Outcome performance indicators compare set objectives with actual results. Areas covered by these indicators include sales, market share, success in opening new accounts, customer satisfaction, loyalty, and profitability. Behavior performance indicators measure both skill-related qualities and activities that are important in fulfilling responsibilities of the sales job: on the one hand, for example, adaptive selling, education, sales-related knowledge, sales experience, product knowledge, intelligence, and empathy; on the other activities such as sales presentations, planning, and support.

Despite the traditional emphasis on the choice facing companies between focusing on either behavior-based or outcome-based control measures, the two can in fact be understood as complementary. This is especially the case for the industrial goods and services industries, where customer relationships, doing business through strategic alliances and collaborations, and coordinated team-based selling play important roles.

Outcome-Based Performance Indicators

Outcome-based performance control resembles a market contracting arrangement in which salespeople are left to achieve results "in their own way" by means of different strategies, processes, and activities. In other words, salespeople are held accountable for their results (outcomes) but not for how they achieve these results (behavior). Historically, sales force performance appraisal systems have emphasized the former rather than the latter, particularly in determining compensation. A major reason for this is the availability of simple, seemingly equitable standards in the form of sales volume or dollars.[159]

Dollar Sales, Sales Unit Volume, Sales Quotas

Partly because of the ease with which orders usually can be linked to the individual salesperson, the dollar is a popular and readily available sales standard. It is the single most cited performance index in published research reports and is very commonly used in practice. Sales-unit volume is another popular standard because of its intuitive appeal as an index of the breadth and depth of sales. Finally, sales quotas have become widely accepted as a convenient, easy-to- provide and measure standard for assessing and controlling the single salesperson's performance; at the same time, they serve as an effective motivation tool.

Though widely used, these simple sales performance indicators have several disadvantages. The first and most obvious is that they do not take account of the cost of the goods and services sold and the costs of selling. This may result in suboptimal gross margins, as salespeople might set prices or determine discounts at too low a level in order to close a deal. In addition, use of these indicators becomes problematic when a considerable amount of time has passed between sales activities and the actual sale. This is especially true in some industrial goods and services industries such as the machine tools and capital goods industries, where a major sales deal involving for example a power plant or a paper machine will take between six months and two years or more to consummate. This can make an incentive system based purely on dollar sales, sales volume, and sales quotas ineffective, since by the time the deal is closed the team may have changed. Furthermore, the motivating effect decreases when the action is detached from the result.

Gross Margin/Net Margin

Other widely-used performance indicators are gross margin and net margin (sales minus the cost of the salesperson). These indicators take the costs of goods and services sold and/or the cost of selling into account. This can reduce the incentive to maximize sales at the expense of margins in cases in which dollar sales, sales volume, or quotas are the only criteria used.

ROAM

ROAM (return on assets managed) takes a capital budgeting approach to evaluating sales force performance. In addition to considering the expenses directly attributable to the sale and the sales force (such as gross margin and net margin), ROAM includes the investment in working capital associated with sales generation. The reason for this is that investment in inventory and accounts receivable is a necessary requirement for and consequence of sales.

Inventory and accounts receivable represent a significant portion of the total assets of most firms; this is particularly the case for industrial goods and services. Return on assets managed is defined as follows:

ROAM = (CD – NS) × (NS – AM)
where
CD = contribution dollars;
NS = net sales;
AM = assets managed (usually inventory and accounts receivable).

Although in most companies production managers are responsible for inventory and financial or credit managers are responsible for accounts receivables, through an investment of effort in particular products within the firm's total product mix and a focus on particular customers and customer segments, the sales force can still have a dramatic impact on the organization's investment in both types of assets.

Residual Income

Residual income is calculated as gross margin minus direct selling expenses and committed working capital, usually accounts receivable costs and inventory costs:

RI = GM – DSE – ARC – ICC
where
RI = residual income;
GM = gross-margin dollars;
DSE = direct selling expenses;
ARC = accounts-receivable cost;
ICC = inventory-carrying cost.

Using residual income as a measure of sales management's performance is consistent with the corporate goal of maximizing shareholder value. It takes into account the potential for profitable growth. As long as the projected residual income for a particular customer is positive, it will add to the company's capital base and thus contribute to shareholder value. Growth objectives are also incorporated because assets managed are directly related to sales. A target level of assets managed may be derived by dividing an organization's total sales objective by an historical or target rate of inventory and accounts-receivable turnover. No matter how fast sales grow, target residual income will not be exceeded unless the target rate of return is greater than the cost of working capital. At the same time, growth objectives are reflected in the target sales volume. As a result, both productivity of assets managed and sales growth are included in the target residual income and are derived directly from overall corporate objectives.[160]

Behavior-based Performance Indicators

Behavior-based performance control systems typically involve considerable monitoring of sales processes and salespeople's activities and capabilities. Furthermore, high levels of management direction of and intervention in the activities of salespeople are used to evaluate and compensate the sales force. Therefore subjective and more complex methods based largely on what salespeople bring to the selling task (e.g., aptitude, product knowledge), their activities (e.g., number of calls), and their sales strategies, rather than sales outcomes, are needed.

The general idea behind behavior-based performance control systems is to focus on performance indicators that an individual can fully control and thus change. A variety of behavior-based performance indicators are reviewed in the following paragraphs. In order to more accurately reflect the different types of behaviors that can be measured in this way, they are divided into sales processes and activities and selling skills.

Sales Processes and Activities (Selling Behavior)

The logic behind measuring how well salespeople adhere to certain sales processes is the assumption that following a predefined *sales process* will result in higher returns than not following it. The individual salesperson is able to optimize his or her results by following set processes. Examples of sales processes are the handling of customer complaints, the preparation and filing of reports, and the planning and implementation of sales calls. For each of these processes, performance standards can be set by management or can be defined jointly between the sales management and sales force.

Measuring selling behavior and activities is based on defining the activities a salesperson needs to perform on a regular basis, monitoring the degree to which such activities are actually performed, and rewarding or penalizing salespeople depending on how well they perform them. Typical sales activities include listening attentively to customers; using established contacts to develop new customers; operating within the budgets set by the company; using expense accounts with integrity; using business gifts and promotional allowances responsibly; carrying out company policies, procedures, and programs for providing information; and submitting required reports on time.[161] In general, activity control is more relevant to managing and directing routine, day-to-day activities of salespeople and aligning these activities with organizational interests. It can be used to ensure that the sales force's activities optimize the chances of both closing a single deal and enhancing the long-term interest of the organization.

Selling-Skills (Capabilities)

Measuring selling skills involves a focus on the development of individual skills and abilities. The underlying assumption for systems based on such attributes is that sales performance is directly linked to employees possessing them on a pre-defined basis. Capability control involves establishing the level of skills and abilities salespeople must possess, monitoring performance in relation to these attributes, providing guidance for improvement if necessary, and rewarding and penalizing employees on the basis of achieved sales skill levels.

Some studies suggest that the effectiveness of the sales force is related to its ability to accurately assess customer beliefs about product performance, its ability to use this assessment to select influence strategies, and its ability to continuously re-evaluate the impact of the influence strategies chosen and adjust strategies when necessary.[162] Other typical selling skills and capabilities include education, sales-related knowledge, sales experience, product knowledge, intelligence, and empathy. In general, capability control is likely to increase a salesperson's intrinsic motivation, selling skills, and relationship with the supervisor. This in turn will often result in increased performance. Capability control is particularly attractive in an environment in which lifetime employment has been replaced by the concept of lifetime employability, as is often the case in highly turbulent environments.

The Role of Context

There is no single performance indicator that will fit in every company and every situation. For this reason, companies combine single performance indicators into complex governance structures. In any case, the moderating role of contextual factors – most centrally environmental turbulence and selling complexity – on the choice of an appropriate control system is even more important than salesperson characteristics; these factors are discussed below.

Environmental Turbulence

Manifestations of environmental turbulence can emerge in frameworks of the market, technology, and competition.[163] In situations marked by high levels of turbulence, in other words rapid, continuous, and simultaneous shifts in the state of contextual factors such as regulation, technology, and value chain configuration, time usually plays a dominant role: with increasing turbulence, the value of information decreases faster. The fast diffusion and wide use of knowledge thus becomes one of the most important factors in managerial decision making. As a result it is advisable for companies facing very high levels of turbulence to encourage a fast diffusion and sharing of knowledge within the sales force, together with an active application of that knowledge in the sales process through the company's performance measurement system. In such situations, behavior-based performance indicators, especially those concerned with capabilities and internal processes, represent a good choice.

In turbulent conditions, organizations need to maintain a great deal of flexibility – the more uncertainty there is, the more frequent the need to alter marketing and sales plans and turn to innovation. This suggests that for the sake of such flexibility, the performance measurement system should guarantee a high degree of freedom to the individual salesperson. As suggested, outcome-based performance indicators usually allow the sales force a higher degree of freedom than those which are behavior based. These, however, can themselves offer a great deal of freedom when focused on capabilities and resources supporting the efficient sharing of knowledge rather than on rigidly defined sales processes.

Selling Complexity

The degree of complexity of the goods and services sold inevitably influences the task of the sales force, hence the appropriate design of the control system. Selling complexity increases when it takes a comparatively long time for customers to make a purchase decision (long selling cycles), several people rather than a single person are involved in the purchase decision, customers require a relatively high level of information, and the buying task is new rather than routine.

Selling complexity has important implications for both sales force activities regarding customers and the design of the appropriate control system. The selling of complex goods and services usually requires high-quality customer service, highly effective interaction with customers, and professional attention to customers from the sales force. These activities will often be delivered by a selling team combining different skills and capabilities. Thus the more complex the goods and services sold, the more important a sales force that is team oriented and highly skilled in sales planning, customer support, and customer service. In addition, there is a need for creative and knowledge-intensive sales strategy. A control system that monitors behavior as well as capabilities can be tailored to meet such a challenge. Necessary behavior such as team selling (for example, measured in the number of joint calls) and taking time for planning, together with other activities designed to support the team approach, can be reinforced. Despite this benefit, a high level of behavior-based control may also lead to standardization of the work processes and thus reduce the level of sales force discretion. Adding outcome-based elements to the control system would then be called for, in order to ensure a certain degree of freedom for the sales force and promote more creative and sometimes knowledge-intensive sales tactics.

Another catalyst of complexity is the customization needed to sell goods and services – always a more complex process because it requires an individualized approach. As organizational control theory suggests, an outcome-based control system is appropriate in a situation of imperfect task programmability.

Designing Sales Force Performance Control Systems

Combining the contextual factors of environmental turbulence and selling complexity results in four settings with different requirements for an appropriate sales performance control system.

Setting 1 represents low selling complexity in a highly turbulent environment. This could be the case with the sale of standardized machines and equipment to the semiconductor industry. In relatively standardized selling processes, task programmability is high, in other words customer needs are more homogeneous and sales force activities fairly programmable. But despite the high task programmability level, the turbulent environment means that outcomes are hard to predict. Especially in the case of long selling cycles, half-yearly or even yearly outcomes can be difficult to measure as deals sometimes take between one and two years and more, with intermediary results not easily at hand. A behavior-oriented control system is most suitable in such a context.

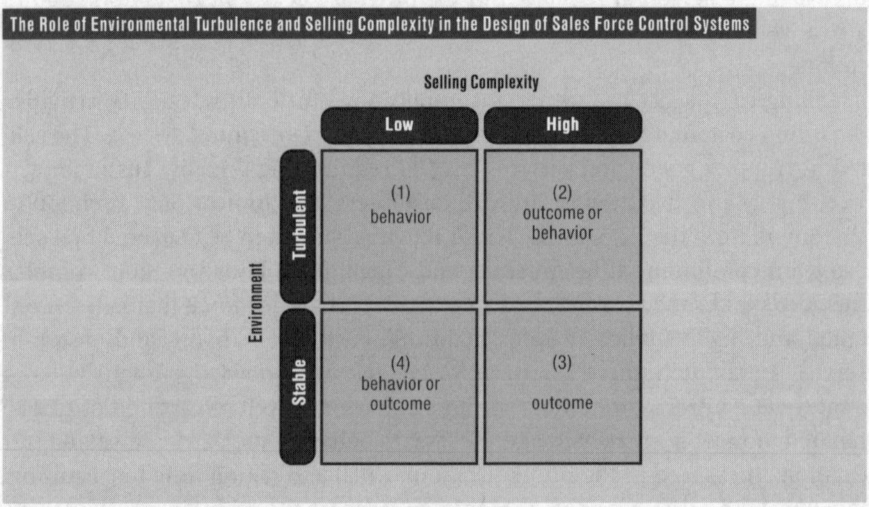

Fig. 1: The Role of Environmental Turbulence and Selling Complexity in the Design of Sales Force Control Systems
Source: Zoltners, et al: Sales Force Design

Setting 2 represents high selling complexity in a highly turbulent environment, for example the sale of customized machinery to the semiconductor industry. Outcomes are not easy to measure in this context, again suggesting the advisability of a more behavior-based system; however, due to selling complexity, task programmability is also rather low, meaning that it is more difficult to clearly define and closely monitor certain behaviors or capabilities, since their influence on the outcome cannot be defined. It is thus best for management to here use a combination of behavior-based and outcome-based elements, with the former group mainly focused on capabilities. The outcome-based elements should focus on profitability (residual income or any other standard that also considers the cost of selling), as this would give the sales force more flexibility on how to achieve its targets.

In setting 3, the sales force is confronted with high selling complexity in a rather stable environment. This is often the case in saturated markets. Here, competition is intense and is driven by process innovation and continuous cost cutting; disruptive technologies and breakthrough innovation play a less important role. Customers might request customized solutions to support their needs; they often shift from buying products to buying services as they start to reduce their vertical integration. In this rather individualized setting, task programmability is low. At the same time, outcome measurability is relatively high due to the stable environment. This is especially the case with many services, which are generally based on long-term contracts and therefore highly predictable.

This situation would suggest use of an outcome-based control system. Especially in highly competitive situations, management should focus on outcome-based indicators that consider the cost of selling and other profitability drivers rather than simpler sales targets like dollar sales, sales volume, and sales quotas. In the case of long selling cycles and thus higher selling complexity, management should either stick to outcome-based indicators through application of some sort of "percentage of completion" concept or else add a small number of behavior-based indicators to account for the deterioration of outcome measurability in such situations.

Setting 4 refers to low selling complexity in a stable environment. An example of this situation is the selling of standardized industrial goods to a firm in a stable market with no major technological disruptions and low to medium competition. As indicated, low selling complexity suggests good task programmability. Knowledge about the buying and selling process is here widely shared; hence the relation between behavior and outcome is predictable and measurable. At the same time, outcome measurability in such stable environments is usually high. Management can thus decide between control systems oriented towards either behavior or outcome, or else opt for a combination of the two: a choice complicated by the fact that the control system itself influences sales force behavior and thus outcome, mainly through its effect on sales force motivation and the cost of the control system.

Conclusion

The measurement of sales force performance needs to be based on a sound understanding of environmental turbulence and the level of selling complexity. One factor to be considered in setting up the correct performance control system is that the close supervision tied to behavior-based systems leads to an increase in motivation levels. Such systems allow organizations to develop positive forms of cognition and capabilities, attitudes, and behavioral strategies. Furthermore, research has noted that the use of behavior-based control is strongly related to sales force selling and non-selling behavior performance.[164] But at the same time, the high level of monitoring is sometimes very cost intensive. In developing a firm's control system, management thus has to balance factors related to both cost and motivation.

Cases

1
Creating a Competitive Edge in Global Markets Case: T-Systems[165]

Christoph Burger
Franziska Frank

When global market space is being redivided, established companies suffer attacks from new competitors; but they may also identify new sources of growth. The case of T-Systems offers a view of an effort to defend national market share while aiming at international growth in a world of global competitiveness.

The challenges in 2005

October 2005: Lothar Pauly had left Siemens Communications and accepted the position as CEO of T-Systems. Prior to the first days in his new office, he had taken part in a series of workshops, had received briefings from his group of experts, and was now considering the main challenges in front of him: How was T-Systems' revenue outside Germany to be increased beyond 15 percent, and how could this German company finally start participating in the globalized market? How could the competitive advantage of T-Systems as a truly integrated information technology and telecommunications firm be leveraged into a company world where "information and telecommunications convergence" was beginning to turn into a household word?

Deutsche Telekom

The story of T-Systems began with a state decree of 1495 by Habsburg Kaiser Maximilian I giving Franz von Taxis the responsibility of building up a postal service system. In 1947 the Deutsche Reichspost was renamed the Deutsche Bundespost, which was privatized in 1995 through the postal reforms resulting in Deutsche Post and Deutsche Telekom. Deutsche Telekom then implemented a four-pillar strategy: T-Mobile, T-Online, T-Systems, and T-Com. In 1996 Deutsche Telekom went public. The official issue price of 14.57 euros for a DT share hit an all time high of 103.59 euros in March 2000 under the leadership of Ron Sommer. In June 2002, however, the price fell below 9 euros. In November 2002, Ron Sommer was succeeded by Kai-Uwe Ricke, who began a process of stabilization, consolidation, and debt reduction.

In 2005 a restructuring process emphasized the increased importance of client orientation: T-Com and T-Online were integrated into one business area

Bringing Technology to Market. Olaf Plötner and Robert E. Spekman

ISBN 978-3-527-50270-7

covering broadband and fixed line services. T-Mobile continued to run the mobile sector, while all of Deutsche Telekom's business clients were bundled into T-Systems.

T-Systems

DT's T-Systems specializes in information and communication technology. The firm was formed in 2000 when DT acquired the IT-Service company debis Systemhaus from Daimler Chrysler. T-Systems has 52,000 employees in two sections: Enterprise Services (ES), with roughly sixty multinational companies as clients, and Business Services (BS), serving 160,000 large and medium-sized enterprises.

ES designs and delivers customized services in IT and communications to large clients. BS concentrates on standardized solutions and products for smaller customers. It also has the overall internal expertise for standardized products in information and communications technology (ICT) services. This twofold structure is reflected in a management board consisting of ten Germans: Lothar Pauly as CEO, two members with overall responsibilities, three representing BS, and four ES. There is a growing focus on industrial segments: within a matrix structure, particularly important segments such as automobile and banking are serviced with tailor-made solutions.

The total revenue in 2005 was 12.9 billion euros, a slight drop from 13 billion euros in 2004. ES contributed 8.4 billion euros and BS 4.5 billion euros. EBIT (earnings before interest and tax) stands at 0.4 billion euros, as compared to 0.5 billion euros the year before.

Growth Strategies: Customers, Products, and Regions

The acquisition of debis had given T-Systems new customers serviced by both IT Solutions and TC Services. In the intervening years T-Systems had managed to gain about 50 percent of the German telecommunications (telco) market and 30 percent of the German IT market. A number of large deals had strengthened the large-business segment. In the medium-size segment, however, a market share that had been about 70 percent had now dropped to 35 percent.

Regarding products, value-added IT services stood next to commodity telecommunication. New areas had also been initiated, such as business process outsourcing (BPO), Internet protocol (IP), and seamless services.

The international strategy of T-Systems had originally come via debis. It was called "string of pearls" and had two elements: If the customer entered a foreign country, a T-Systems office (a "pearl") would be established to serve this customer. In addition, the office would try to generate local sales. At the same time, if an interesting opportunity arose, a market could be entered through acquisition of a medium-sized company.

Having taken this growth path, by 2005 T-Systems had twenty-three offices worldwide; it had near-shore and off-shore locations in the vicinity of India, Hungary, and Slovenia. In Austria and Switzerland, local revenue placed the company among the top ten corporate earners. In all other countries, growth remained sporadic. Although a number of organizational models had been tested, no ideal set-up had been discovered. There was also need for a set of internal regulations. Georg Jakob (head of the commercial management division of M&A Enterprise Services of T-Systems) described the situation as follows: "There is limited consistency in our international activities so far. Branches have often acted and interacted with the rest of the T-Systems organization according to local strategic targets and idiosyncrasies."

The ICT Market

ICT is the business of developing and using technology to process information and aid communications. Communications technology consists of products such as voice-over IP, datacom, and communications equipment. Apart from IP and added-value services, it is a commodity market where differentiation is hard to achieve.

Market Estimate Telecommunications

Global Market Estimate in bn €		2003	ca. CAGR 2003-2007
TC Infrastructure Services	LAN solutions and outsourcing	47	5%
	Voice equipment incl. services	24	4%
TC Services	Business voice	172	-2%
	Incl. services	76	-5%
	IP solutions	38	13%
	Mobile solutions	13	15%
	Total	**369**	**1%**

Table 1: Market Estimate Telecommunications
CAGR = Compound Annual Growth Rate

IT services and products – systems integration (SI), business process outsourcing (BPO), and information technology outsourcing (ITO) – have high value added. Each market obeys its own rules: whereas due to its localized content SI is predominantly a local market, BPO and ITO are both businesses involving economies of scale and have thus become global. The market is growing worldwide and is estimated to have reached 680 billion euros by 2007.

Due to the complex nature of the ICT market, competitors attack from different sides. Some originate in the world of IT (IBM), others in telecommunica-

Market Estimate IT Markets

SI, BPO, and ITO, Market Estimate bn. €					
Country	2003	ca. CAGR 2003-2007	Country	2003	ca. CAGR 2003-2007
			Czech Rep	0.4	4%
Australia	10	7%	Hungary	0.5	3%
China	3	11%	Poland	1	3%
HongKong	1	3%	Russia	1	3%
India	2	22%	Rest of EE	1	1%
Indonesia	0.2	6%	**Sum EE**	**4**	**2%**
Japan	73	4%	Austria	3	1%
Korea	5	9%	Belgium	5	1%
Malaysia	1	6%	Denmark	3	4%
New Zealand	1	9%	Finland	3	2%
Singapore	2	8%	France	22	2%
Taiwan	1	6%	Germany	32	2%
Thailand	0.4	13%	Greece	0.4	1%
Rest Asia Pacific	1	6%	Ireland	1	2%
Sum Asia Pacific	**101**	**5%**	Italy	13	1%
Canada	16	6%	Netherlands	11	2%
USA	252	5%	Norway	3	2%
Sum NA	**268**	**5%**	Portugal	1	4%
Sum LA	**18**	**9%**	Spain	8	4%
Israel	1	2%	Sweden	6	3%
MEA	2	4%	Switzerland	6	3%
Africa	3	5%	UK	53	7%
Turkey	0.4	5%	Rest of WE	1	0%
Sum MEA	**6**	**4%**	**Sum WE**	**167**	**4%**

Table 2: Market Estimate IT Markets

tions (BT Global Services), and yet others in consulting (Accenture). Marked by strong price pressure, recent years have seen considerable consolidation: high levels of industrial standardization and increasingly comparable products and services. Mergers and alliances are a preferred tool for establishing a strong market position.

International customers are looking for ICT providers who wish to build a long-term global partnership.[166] A survey[167] has shown that regardless of size, more than 60 percent of all firms either have or plan a single-sourcing system for IT, more and more frequently including TC. They expect leverage of economies of scale and want to be treated as global players.

Brief Look at Competitors													
Companys		Revenues in bn €			Product Split				Country Split				
	# Empl. (.000)	2003	2004	2005 e	Outsourcing	SI	Consulting	Telco	Americas	Key country	Europe, Middle East Africa	Asia Pacific	Rest
Accenture	115	10.4	11.0	12.1	37%	0%	63%		45%		48%	7%	
BT GS	25	8.6	8.4	9.4		12%		88%	1%	UK 92%	6%	1%	
CSC	80	12.0	13.1	12.1	45%	55%			60%	UK 15%	16%		9%
EDS	117	18.2	16.6	15.8	88%		12%		51%	UK 16%			33%
HP Services	69	10.9	11.1	12.2	18%	64%	18%		36%		42%	22%	
IBM GS	215	37.7	37.3	37.8	41%	28%	31%		52%		26%	15%	7%
SBS	39	5.0	4.7	5.4	46%	26%	28%		8%	Ger 43%	44%		5%

_ Figures for geographical split for all of BT

Table 3: Brief Look at Competitors

The Plans for T-Systems

After his discussion with Wilfried Peters (CFO T-Systems), Lothar Pauly reflected on the context of the internationalization strategy: How could the firm leverage its competitive advantage in a globalized market? A number of relevant issues – regarding regions and customers, products, and the organization's nature – had been raised both in workshops Lothar Pauly had attended and later. Some of the salient observations continued to ring in his mind:

- "If T-Systems wants to grow, it can really do so only outside Germany. With its 13 billion euro revenue it covers a large part of the whole ICT market in Germany." (Katrin Horstmann, board member responsible for strategy)
- "It is difficult to communicate that there is to be a focus on a few countries only, and then we go on building up staff in India, Slovakia, and Hungary. We must rise to that challenge now and explain – if we don't then in a number of years it will backfire." (Herbert Schaaff, member of the managing board, responsible for human resources)
- "To be successful internationally we should build up additional industry expertise." (member of ES)
- We hope that there will be no limitation on the "opportunities in the emerging markets of Asia and especially larger English speaking markets like the USA, other parts of Europe, and Australia." (Milind Kaulgud, head of the human resources department, T-Systems India)
- "The business of IT is interesting as long and as soon as it aids the sale of TC products and services." (Deutsche Telekom CEO Kai-Uwe Ricke)
- "It [the ICT-market] is a result of technological change; since the future is internet protocol, the chief technology or operations officer will quite rightly also be the person in charge of telecommunications procurement." (Ralf Nejedl, executive vice president, Country Operations Western Europe)

- "We belong to and compete against the big guys in the global markets. Our next battlefields can be anywhere in the world. It is thus vital to employ local CEOs and not German ones in our key countries." (Michael Heuser, executive vice president for people development and culture)

The Challenges Addressed in 2006

As of July 2006 Lothar Pauly has become convinced that in a globalized market such as ICT, the firm's strategy shift will enable it to leverage the competitive advantage inherent in being a truly integrated ICT house. He sums up the key elements of the new strategy as follows:

Regions

1. The German market does not offer sufficient growth potential. International expansion is a must if T-Systems wishes to grow in the long run.
2. T-Systems has to be able to realize perfect worldwide delivery for any client yet concentrate its sales focus on defined growth markets.
3. Western Europe, with a growth rate of 3 to 5 percent per year, is the main region where T-Systems should be engaged. The competitive situation is strong but offers enough opportunities, in particular for a Europe-based company. The American ICT market is large and growing but has many strong competitors. Eastern Europe, the Asia-Pacific region, and the Middle East are still very small markets, so that an engagement can only be of strategic value – except for Japan, where the competition is simply too strong.
4. Market size is clearly important in horizontal product lines such as SI, BPO, and Desktop Services. Large deals can only be acquired if there are economies of scale in a particular country or region. Thus, market-position goals need to be set for each country. With regards to industry know-how, market position is less relevant in the local markets than over the whole industry.
5. In view of the relevant market data, the largest European countries are most suited for expansion of the firm's activities. The aim is to be among the top five ICT firms within three years, apart from the highly competitive United Kingdom market, where the goal is the top ten. This has already been achieved in Austria, Switzerland, and – through acquisition of gedas – in Spain. (With 5,500 employees, the former Volkswagen IT house gedas has a considerable international focus. In 2004 revenue was 567 million euros split as follows: Germany 28 percent, Spain 23 percent, USA 11 percent, UK 10 percent, France 10 percent, Mexico 9 percent, Brazil 5 percent, the Czech Republic, Argentina and China, 1 percent each.)

6. Although it is clear that quick growth for T-Systems cannot come from the emerging markets, a presence there is vital in order to be in place when the markets mature. The first priority of these offices is to service the large international clients from the core markets needing services in these countries.
7. Due to the focus on a few countries, these can allocate more resources, which in turn should speed up their organic growth. For non-organic growth T-Systems can pursue one of two options, acquiring either additional captive IT-subsidiaries such as gedas (bringing new large contracts with them) or medium-sized independent IT service providers.

Products

1. Although the market for some telco products such as IP and value-added services with high differentiation potential has been growing, many telco products such as Legacy have low value added in a largely price sensitive market. In most countries, the market is dominated by large telecommunications companies committed to defending their market share. International expansion needs to come mainly from the high-value-added IT Services, a domain in which T-Systems can differentiate itself in a heterogeneous market. Telco products can then be sold as a follow-on product.
2. When it comes to big deals, industry experience is a key success factor. With the acquisition of gedas at the end of 2005, T-Systems strengthened its automotive experience in offering end-to-end solutions to numerous car manufacturers (DaimlerChrysler, VW, BMW, and Ford).
3. Before the strategic shift, individual countries decided what product and services they wished to offer. This does not suit a company that aims at serving global markets. For this reason a standardization process has been initiated, its goal being both the availability of outsourcing and telecommunication in all countries and the emergence of products and services with higher value added.

Customers

1. In order to reach the above-mentioned goal of being among the top five ICT firms in each of the key countries (among the top ten in the UK), and in order to generate the necessary revenue in an acceptable time frame, a policy of serving only larger clients is vital.
2. Worldwide sales outside the focus countries are only pursued for *projects*. An example is the FELIS project (*Flexibles Einsatzleitsystem Innere Sicherheit*), a computer-aided system for managing emergency calls. If a "non-focus" country expresses interest, a sale to such a country can take place if

about 80 percent of deliverables are a replication, the project is limited in time, and a project team is able to cover it without building up the local staff.

3. A worldwide customer relationship management system (CRM) is in place, moving T-Systems one step closer to true key account management. In the French automobile sector, hiring new personnel with client and industry knowledge and creating a new process for managing key accounts has already resulted in new contracts.

Organization

1. Since there is a sales concentration on core countries, yet the company is acting in a global market, worldwide delivery is only possible when it is split from sales. The delivery unit is thus structured like a cost center and has the duties of ensuring smooth delivery and, beyond that, optimizing the global-production value chain.
2. Due to their importance, the CEOs in the various focus countries are moved up in the company hierarchy. The focus countries report directly to Axel Knobe (head of Sales & Services Management at T-Systems Enterprise Services), and the other countries have one intermediary level added.
3. When countries are part of one particular industry line such as finance, experience has shown that other lines are not serviced as effectively in that country. It is thus important that the CEOs in focus countries run their business as profit centers, but also have responsibility for all the industry lines. To maximize the potential of countries and industry lines, focus-country CEOs and industry lines work closely together in so-called joint management teams, with the CEOs discussing important developments and formulating strategies with the heads of the industry lines.
4. A number of complex issues have to be addressed in order to balance an organization of focus-country profit centers and delivery cost centers. Markus Wagener from HQ Corporate Business Architecture integrated around fifty people from company headquarters, every service unit, and selected lines of business units into one project to define the governance of the new organization. The result is a ninety-page guideline covering all aspects of international cooperation: international strategy, roles and responsibilities in sales and delivery, cross-border contracting, legal and tax requirements, and contracting and financial regulations.
5. To balance coherence and independence, both the CEOs of the key countries and human resources managers will be local hires, while the chief financial officers will still be German. CEOs will meet every month, chief human resources staff every two months.

Conclusion

In retrospect, many aspects of the new international strategy appear to have been very clear-cut. But Lothar Pauly recalls the high input of leadership and management it took at all levels to ensure that existing structural issues could be overcome in favor of strategic considerations. At present, the important leadership challenge is to maintain momentum: to regularly adapt and strengthen T-Systems' position in a fast-changing market.

Interview with Lothar Pauly[168]

Franziska Frank

Mr. Pauly, is T-Systems an international company?

We're an international company, but we still have to become somewhat more international. On the one hand, more and more of our customers are acting globally and expect internationally operating service providers. On the other hand, we can only reach our growth targets if we succeed in gaining customers and market share outside Germany.

What are the concrete goals of your international strategy?

It allows us to strengthen our European market position, and it further enlarges worldwide supply capacity for our multinational customers. At the same time we're trying to achieve a leading worldwide market position in selected industry segments such as the automotive segment.

In acquiring local customers, our focus has been on countries in which we see the largest market potential for us as T-Systems. We've defined Western European countries according to this criterion–which is to say France, Italy, Spain, Great Britain, Austria, Switzerland, Belgium, and the Netherlands. It's our goal to be among the top five providers in these countries within the next three years – in Great Britain among the top ten.

Looking back, what are the three most important lessons you've learned?

My years abroad were certainly formative for me as a manager. Especially the pressure of the fast-growing Asian and Eastern European markets encourage and require a management style marked by quick thinking and acting.

Another insight stemming from that period is that for on-site success, involving local managers is a sine qua non. Anchoring and networking local CEOs in their own countries will also be a decisive gain for T-Systems.

Another crucial factor is that a company's success will depend on customer processes to a stronger degree than is the case at present. At the customer interface, qualities such as additional services and quick reaction time will be decisive. To this extent, what is at stake is implementing solid and excellently managed key account management.

What have been the three most important lessons learned in respect to the new international orientation of T-Systems?

In the restructuring process the important thing was integrating all relevant employees and local companies early on. Especially the creation of interna-

Bringing Technology to Market. Olaf Plötner and Robert E. Spekman

ISBN 978-3-527-50270-7

tional governance had enormous significance for T-Systems as a firm with a complex cultural background. With this project as a basis, the employees could not only internalize the new strategy but a common culture as well.

This process generated growing awareness of the internationality of T-Systems – along with increased recognition that the company still needs to become significantly more international. That's our greatest challenge: T-Systems has to adjust itself more strongly to this international market, not only in an organizational sense but also mentally. It's only then that we'll be able to accomplish the same thing outside Germany that we accomplished within the country, namely overtaking competitors (in this case IBM).

What is your sense of the challenges facing T-Systems in the global market over the next ten years?

As far as the market is concerned, consolidation will continue. In the future, IT subsidiaries will still be taken over by bigger players. And here one or another larger company will have constant problems running an IT service business driven by economies of scale with consistently high intensity.

Admittedly, in view of the breadth of the theme and the low entry hurdles, we'll continue to see niche providers. For an international ICT service provider like T-Systems, this means integrating specialists as partners and freeing the customer of the burden of having to negotiate with every single player.

2
Creating New Business by Redefining the Value Chain
Case: MAN Roland Druckmaschinen AG

Bülent Gögdün
Olaf Plötner

When speaking about the newspaper *El Diario,*[169] satisfaction at MAN Roland Druckmaschinen AG could not have been more profound. At a ceremony on May 2, 2006, Gerd Finkbeiner, chairman of the executive board of MAN Roland Druckmaschinen AG, and Peter Kuisle, executive vice president of sales, had signed a very important contract with the management of the newspaper. The contract was both for the delivery of four new systems amounting in value to 54 million euros and for after-sales service amounting to 15 million euros. During the reception, the customer had also made a presentation in which he referred to the "dawn of a new relationship" and emphasized the significance of the partnership between *El Diario* and MAN Roland.

MAN Roland

MAN Roland Druckmaschinen AG, a company founded in 1845, develops and manufactures web offset systems[170] for printing newspapers and advertising material as well as sheet-fed offset systems[171] for printing published material, advertising, and packaging. The company also offers extensive services and software solutions for the printing industry: activities including the planning, construction, and commissioning of complete printing plants and after-sales services such as the delivery of spare parts, inspections, and overhauls of printing presses.

The printing press industry was long dominated by German manufacturers. With a turnover of 3,360 million euros in 2004-2005, Heidelberger Druckmaschinen was the largest supplier on the market, followed by MAN Roland (1,783 million euros in 2005) and KBA, Koenig & Bauer (1,621 million euros in 2005). In 2005, a number of other suppliers from various countries such as the United States, Switzerland, and Japan accounted for the remaining 35 to 40 percent of the market. MAN Roland was the world's leading supplier of web offset systems and the second largest manufacturer of sheet-fed offset systems. In 2005 the company had some 8,800 employees. The main facilities were located in Offenbach and Augsburg as well as in Plauen, where both the systems themselves and the spare parts not provided by suppliers were manufactured. The sales network included some 120 branches and sales partners worldwide. A service network was being developed, making growing use of

Bringing Technology to Market. Olaf Plötner and Robert E. Spekman

ISBN 978-3-527-50270-7

tele-support centers, providing immediate support by on-site technicians in the checking and if necessary repair of the systems.

With an export share of 83 percent in 2005, MAN Roland achieved an operating result of 65 million euros. This reflected considerable improvements in both web-fed and sheet-fed offset systems. The sheet-fed offset systems division, which had been running a loss for three years, managed to achieve a turnaround and post a positive result (+6 million euros as opposed to -41 million euros in 2004). The web offset systems division reached a new high in terms of order intake with 1,183 million euros and increased its result to 59 million euros (44 million euros in the previous year).

The Printing Industry

Even though 2005 was a successful year for MAN Roland and a further increase in the operating result is expected for 2006, the annual report for 2005 contained the following statement: "The situation in the printing industry remains tense worldwide. The sector has been characterized by continuing concentration, plus globally fierce competition between printing press manufacturers." The crisis of recent years had not yet been forgotten. The seriousness of the situation for printing press manufacturers became clear in a remark by Bernhard Schreier, CEO of Heidelberger Druckmaschinen, in July 2003: "I can choose between discounts for the customer and a redundancy plan for the workforce."[172] In 2005 things gradually started to turn around for the printing industry. Experts expected the global printing markets to grow by 2 to 5 percent per year. This growth was to be generated by replacement investments in industrialized countries and in particular by additional volumes in the populous emerging regions such as Eastern Europe, South America, China, and India. Strong economic growth, growing populations, increasing prosperity, and improving educational standards were making them all attractive future markets. In spite of the fact that the advertising and printing industry was recovering, the competition remained fierce. The prices in the web offset systems business were up to 20 percent lower than they were a decade ago.

El Diario

El Diario was considered the flagship of the media corporation Victor Venegas S.A. Victor Venegas was one of the largest media corporations in Latin and Central America, with an estimated 220 million dollars turnover in 2005. The majority of the company was family owned. The company owned newspapers, magazines, TV and radio programs, Internet portals, and book and music publishers. Founded in 1912, *El Diario* was the second-largest Spanish-language daily paper in the world in terms of circulation in 2005. It had a print run of over 1.9 million copies per day, was published in seven editions and several countries, and had set itself ambitious growth targets.

El Diario and MAN Roland have had a good business relationship existing on several levels for many years. Over the previous ten years, the paper had ordered an average of one system per year from MAN Roland. In spite of this, MAN Roland still had to compete fiercely for orders with competitors such as König & Bauer, Goss, WIFAG, TKS, and Mitsubishi.

The Project

At the beginning of 2004, *El Diario* asked several manufacturers of web offset systems to submit quotations for three different packages: the first package for one printing press, the second for four, and the third for six. In doing this, the new purchasing manager, who came from a renowned international consumer goods company, was breaking new ground. A decision pyramid was to be used to find the best bid. The manager laid down five criteria: basic requirements, quality, service and support, costs, and innovation. In evaluating the quality of suppliers, *El Diario* not only gathered detailed information about the quality characteristics of the products but also about the quality management carried out at the manufacturers. The prices for the new systems were submitted at a transparent price event. This meant that the suppliers had to submit their prices anonymously but visibly for all parties per Internet and had the opportunity to optimize their bid by a certain deadline.

It was assumed that the paper did not wish to purchase all of the packages simultaneously; however, it was not clear to the bidders to what extent it was interested in purchasing which package. MAN Roland was, as usual, not the lowest bidder for any of the packages. In addition, the purchasing department informed MAN Roland that the proprietor family wished to avoid any dependency on one supplier, and that this was particularly true in the case of MAN Roland, since they had already supplied most of the existing system fleet.

Despite this situation, MAN Roland was able to win the order. The main reason for this was that in a difference with normal business procedure, the company offered extensive services based on a fixed price for a period of twelve years. This had not been stipulated in the invitation to bid, but it was a convincing argument for the customer, since it would make it possible to budget for the life-cycle costs of the printing systems.

For this reason, the bid also included an offer of services for three further MAN systems already operating at *El Diario*. On July 27, 2004, the letter of intent was signed for four new systems and the new service package. However, almost two years of negotiations would follow before the contract could be completed. The service agreement needed to include a definition of what service was to be provided when and at what price. In this framework, Eduard Hoffmann, the new head of the service division at MAN Roland, observed that "even when it is possible to cover many eventualities, in the end it always remains a matter of trust."[173] After several days of negotiations with lawyers and administrators shortly before the contract was to be closed, a number of issues were still open. It was only after a one-to-one talk between a MAN Roland manager and the technical manager of *El Diario* – the two men knew each other well – that an agreement could be reached.

The contract included the delivery and commissioning of the four new systems, valued, as mentioned, at 54 million euros. In addition, extensive services for these systems as well as the three already operational MAN printing presses were purchased for 15 million euros. The service package included the following: delivery of all spare parts on a just-in-time basis; part types to be defined exactly; replacement parts due to accidents, errors of use, and wear and tear not included; inspections once a year; check of bearer rings every two years;[174] overhauls, including a check of the folding unit, every six years (including replacement of parts).[175]

In contrast to larger repairs, both smaller repairs and the cleaning of the systems were to be carried out by the customer. As indicated, the period of the service contract for the four new printing systems was twelve years from the date of commissioning. The three MAN Roland printing systems already in operation would likewise be serviced for twelve years from the date of their commissioning (in 2012 and 2015). In the past, the customer's own service team had carried out a large part of the maintenance and repair work together with inspections. Here, after the two-year guarantee period, the necessary spare parts would often be purchased by *El Diario* at very reasonable prices from smaller Central American suppliers able to construct them to the same design. The only services the customer would purchase from the systems manufacturer were those necessitating a high level of expertise or spare-part complexity. Peter Kuisle estimated that without the new agreement, MAN Roland would only have been able to generate one-third of the service business with the customer. On the other hand, this entailed considerable compromises on price, reducing the normally high profit margins in the service business. However, two important suppliers, the drive-train manufacturer and the bearings producer, have plants in the region, which reduces logistics costs, at least for some parts.

This kind of long-term, fixed-price service agreement is unusual in MAN Roland's particular industry. It has thus represented a new business model for both the customer and the company.

Interview with Peter Kuisle[176]

Bülent Gögdün
Olaf Plötner

Mr. Kuisle, in signing a contract with the company we're referring to as *El Diario*, both your company and the entire industry were stepping onto new terrain with regard to service arrangements. Why were you willing to take this step?

In the first place we're convinced that with this agreement we'll earn money. Beyond that, we've entered a new, highly promising dimension when it comes to working together with customers. And finally, it represents an important initiative for the company's strategic development.

What do you find strategically attractive in taking over value creation processes that until now were generated by the customer?

In our main business, the manufacture of printing machines for newspaper presses, we're active in a branch with low growth rates. We more or less compensate for increased circulation through increased productivity by our machines. For this reason we can only sell machines as a replacement investment or as what is conveyed to us as a technically new request from the customers of our customers, for example an increase in color content or a trend towards qualitatively higher print products. In this way we can achieve yearly worldwide growth rates of no more than 3 to 5 percent. But in upgrading our service business we see a chance to create significantly more sales. At present this sector comprises around 14 percent of our business; but we hope to soon generate 30 percent of our sales here. This of course with a strengthening of our machine business and an upgrading of its market share.

Why are customers interested in your service?

We have learned that on average customers only use around 70 to 80 percent of our machines' potential. The machines could be run at higher net speeds, the setting-up time shortened, and the production possibilities multiplied. We know how to manage this best. We like to speak here of the "three S's," namely service, spares, and supplies. If using our services means optimizing customer processes, sensibly maintaining the machines, and making use of proper replacement parts, then customers can use the machines more efficiently–and that is surely worth the money we're paid.

Then couldn't you actually operate your own machines?

We discussed this once already around a decade ago, but at the time the market wasn't yet ready for such changes. Now we'll carefully examine if an exten-

Bringing Technology to Market. Olaf Plötner and Robert E. Spekman

ISBN 978-3-527-50270-7

sive reorganization of value-added chains is underway in the various branches. Maybe we and our customer were the first to move significantly in this direction.

Why did you take the new direction with precisely this customer?

A new business is always tied to risks. While it's the case that *El Diario* isn't really located on our doorstep, we've known this customer for years and have a trusting working relationship. We know that the company operates its business very professionally and handles the machines properly. That has great importance with regard to the agreement we've made here. At the same time the company represents an important benchmark for the entire newspaper industry. Against the backdrop of our strategic planning, we thus consider it especially valuable to gain a more detailed view of the firm through the closer collaboration.

When it comes to your relationship with the customer, what has changed since taking on your new responsibilities?

The relationship has become more intensive. We're no longer sporadically with the customer for one or another maintenance or repair task, but constantly present for the entire life cycle of the machines. In that situation you not only become better acquainted on a human level, but employees also gain a better understanding of how the other company functions. I should also mention that we've introduced a steering board so that management regularly exchanges information about business developments in general and our collaboration in particular. In this way we've gained a better sense of where things stand with the customer and are more quickly informed about the company's present and future needs. An additional side effect is that we've gained a better view of the performance of our machines over their entire life cycle, and this has given us a decisive push in developing our new technologies. Naturally we've also become more transparent for the customer. For example, the company has a better understanding of our price structure now and certainly does its best to use this knowledge to its own advantage. But altogether both sides profit from the close collaboration. One fact makes this especially clear: earlier, the negotiation process for new machines from the bid to contract signing lasted half a year; with the last contract, everything was already taken care of within four to six weeks.

What effect has the new quality of customer relations had on the sales sector?

Sales managers already need to know a great deal. They have to know a lot about the customer's business – about machine building and the retail market for machines, about working out contracts and different financing models. In long-term service contracts with many unknown factors, above all risk assessment and controlling take on an increasingly important role. Finally, when the twelve years are over we'd like to look back at a generally profitable business –

this means calculating and negotiating differently. Realizing such a goal involves a far better understanding of the customer's business than was previously the case. You have to be able to analyze and evaluate customer processes and especially know the behavioral mechanisms in each of the sales markets.

How do you learn more about the customer's business?

Beyond our everyday customer contact, we systematically try at MAN Roland to better understand our customers' strategic challenges, the trends manifest in their markets. We're thus for instance one of the five strategic partners in the International Association of Newspapers and support the project "Shaping the Future of the Newspaper" – not only financially but above all with substantive contributions. Another example is that together with the "European Web Offset Association" we've carried out a study of the desires and needs of buyers of printing articles. I should also mention that for years Gerd Finkbeiner, the chairman of the board at MAN Roland, has given a lecture series on the printing and media industry at a well-known university, thus already developing contacts with managers before they've begun their careers with potential clients.

Doesn't that simply amount to sales networking?

Naturally, contacts are used in this branch for sales activities. It's nevertheless important for us to be a competent interlocutor for our customers with regard to their own business. Allow me to offer an example: together with a German newspaper publisher we managed to print advertisements on transparent paper with standard newspaper-printing machines. No one had thought this was possible. Under the rubric "applied innovations," this new possibility was presented to various publishers and advertisers, which led to a large American newspaper group becoming interested in the innovation, then profitably selling and producing it to their advertising customers.

What have been the lessons learned following initial experiences with service activities that have been new for your company?

While the contractual arrangements were being made we were sometimes on completely new territory. Although I'm satisfied with the results I still think we might have gathered more input from other industries.

It certainly also makes sense to arrange as soon as possible for support in your own house: when something new is initiated, there are always skeptics in your own ranks. It's then a good idea to obtain a timely commitment from just those employees who will be affected by the innovations.

What's proved its worth was the agreement to keep everything relatively simple, despite its complexity: a variety of machine types with various operating times and service needs – and many unknown factors. Things which are too complicated won't be understood; things no one understands won't be accepted.

But can't conflicts quickly emerge from points that aren't agreed on?

When there's a bad relationship with the customer, conflicts develop even with the most detailed contract. When the relationship is marked by trust, these points can be solved between partners.

3
Value Selling
Case: Leistritz Extrusion

Mario Rese
Richard Steiner

The Company

Leistritz Extrusionstechnik GmbH is a German subsidiary of Leistritz AG in Nuremberg, Germany. For over one hundred years, the company has been a key technology partner providing international customers all over the world with its mechanical engineering products.

Leistritz products are used in the aviation and space industries, in power-plant engineering, the petrochemical industry, automotive assembly, chemical engineering, and the pharmaceutical industry. This explains why some very high-profile companies are among its satisfied customers. Leistritz currently employs about 2,000 staff in its ten subsidiary companies. The products of Leistritz Extrusionstechnik GmbH are primarily used in plastics processing. The core product of Leistritz Extrusionstechnik GmbH, the extruder, is mainly used to process thermoplastic materials. Put simply, plastics are melted down in an extruder (usually in the form of pellets) for the purpose of fusing them with other materials, such as colorants or anti-inflammable substances. The end result is once again plastic pellets, but with modified properties. These serve as the input material for such machinery as the injection-molding machines that manufacture a variety of different products like PET bottles and vehicle bumpers.

The main buyers of extrusion machines are so-called custom molders. They create value by selecting raw materials, including recipes (referred to in the industry as formulations), developing them, and refining the process physically and chemically to maintain consistent product quality throughout the processing cycle. These custom molders make up the core clientele of Leistritz Extrusionstechnik GmbH. The aim of Leistritz is to advise these customers on the design of the extruder and the entire processing plant in order to ensure maximum customer benefit. To this end, plant sales are accompanied by intensive product-related services: materials processing advice, plant engineering, and programming of the application software. Depending on the customer's level of expertise, this may include comprehensive process qualification for such purposes as pharmaceutical production.

Bringing Technology to Market. Olaf Plötner and Robert E. Spekman

ISBN 978-3-527-50270-7

Sector Trends and Structure

The origins of the mechanical engineering segment known as extrusion technology lie in Germany, which has always been one of world's key chemical engineering nations. Consequently, all direct competitors in the extrusion market were initially also based in Germany. Following the successful worldwide marketing of mass-produced plastics, the post-1945 plastics processing industry expanded among the world's key industrial nations. This led to the production of extruders in the larger international markets. The reason for the initial success of these products was the cost benefit of using them in technically less sophisticated plants, but other market-access barriers for established suppliers and cultural influences such as language and service conventions also played a part. The new players learned very quickly and became increasingly adept at using the technical advances of leading German machinery and plant engineering firms. The technological edge Germany enjoyed was thus rapidly eroded.

Today the market for extruders can be broadly divided into three regional zones: Europe, North America, and Asia. Apart from leading international chemical companies, there are many major buyers in these three main regions. However, in the intervening years, many smaller processing firms also became established. One reason for this was the rapidly growing market. Another was the change in customer requirements coming from end producers. Their demand for pellets became more and more diversified, which was reflected in a host of different formulations, and this was accompanied by a preference for smaller and smaller lot sizes. In the competitive custom-molding market, the smaller, so-called local compounders proved to be more flexible and responsive to these customer requests than the big raw material suppliers. As a result, smaller-scale production contracts in particular began to be increasingly awarded to these local compounders. This group of customers is characterized by a mixture of curiosity and skepticism. They are basically interested in any technological innovations that will help them maintain their competitive edge over the larger raw materials suppliers. But at the same time, they are initially skeptical of any new advances and first have to be convinced of the ensuing benefits.

Merchant compounders make up a third group likewise involved in the sale of extruders. For historical reasons, industrial chemical companies have maintained a large distribution network with a wealth of dealers in every imaginable region of the world. These distributors were quick to recognize the potential added value of plastic products. When the chemical industry began to award smaller-scale contracts to local compounders, many distributors were quick to obtain licenses to produce such compounds themselves.

There are now a multitude of custom molders of many different sizes and origins, which has led to fierce price competition. For this reason, almost all custom molders try to create customer-specific developments or formulations of their own. These are designed to offer end producers product-related or pro-

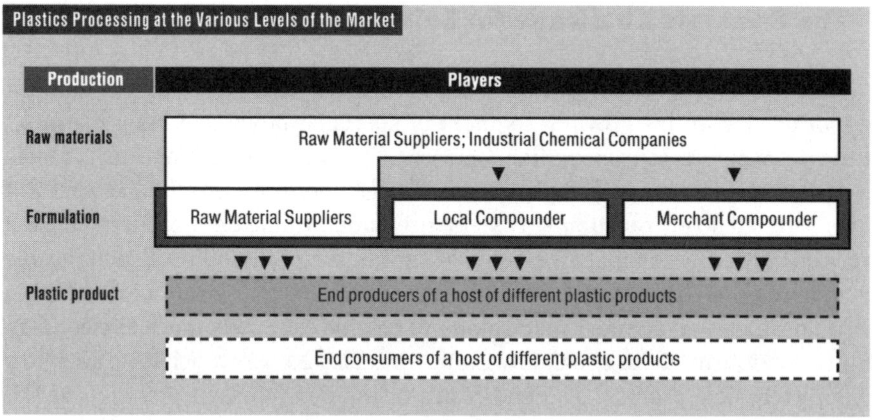

Fig. 1: Plastics Processing at the Various Levels of the Market

cessing benefits. The aim is to circumvent the pressure on prices. In addressing the demand for extruders, this means achieving a balance between economical yet highly flexible production systems. The following illustration summarizes the market situation that has just been described. The section highlighted denotes the custom molders and thus the market for extruders.

Customer-Needs Analysis

An analysis of the situation of local compounders pointed to some very different requirements for the extruders of the future. On the one hand there is the mass producer segment processing simple plastics–or commodities–on so-called single-purpose plants. Economically, single-purpose extruders are generally only required to be affordable and have the highest possible throughput per hour. Technological advances tend to be quite minor in this segment, and there is less potential for differentiation of suppliers. The processes are generally predetermined and recommendations from satisfied customers define the business. The willingness to swap from one supplier to another, despite the high degree of standardization, is not very prevalent, since experienced operations managers are skilled at estimating the costs of such changes. Nevertheless, price does play a significant role in the case of these standard plants.

On the other hand, there is a growing segment of compounders who focus on producing customized formulations. These have to be delivered in increasingly smaller batch lots, JIT (just in time), and to a consistent quality standard. This group tends to require extruders that are small in terms of throughput per hour, simple to clean, and easy to adapt.

The Strategic Challenge for Leistritz

It became clear to Leistritz that the company would also have to look for opportunities in this growing segment of specialist producers. As a manufacturer in a high-cost country like Germany, it could not continue to compete over the long term with suppliers in low-wage countries and the Far East if it concentrated solely on supplying standard plants. In the development phase it very quickly emerged that an essential factor in the profitability of such flexible plants was the changeover or refitting time. This led to a vision of developing a multi-functional extruder that would not only be extremely quick to clean – in up to one-fourth of the usual downtime – but also equally quick to convert to a completely new process. The end result of the development process was the Leistritz Flexxline (fig. 2). By using two processing units that can be exchanged within a very short space of time, the desired reduction in changeover time compared to that of standard extruders was achieved.

Fig. 2: Flexxline Extruder by Leistritz

However, it was still clear to responsible parties that market success was by no means assured. The problem was that the Flexxline extruder was much more expensive to produce than most comparatively sized single-purpose extruders already on the market, due to the two processing units involved. The assessment criteria compounders had been using for many years to weigh up the merits of a new plant came down to two points: price and throughput per hour. On the first point at least, the Flexxline was clearly inferior to standard extruders. In addition, the benefit of flexibility did not seem to resonate well with the sector. This meant it would be no small challenge to fundamentally change the way the industry evaluated the merits of various extruders.

It was a matter of not only explaining the benefits of flexibility to the compounders from a technological perspective, but also making the economy of such a concept tangible. The sector, in other words, had to become aware of the very principle Leistritz itself had grasped that the key success factors were not plant cost or throughput per hour but rather the changeover or refitting times.

In view of this challenge, Leistritz developed their concept of providing integrated technical and financial advice to customers. As well as giving technical explanations, equal weight was placed on solid financial arguments and maximizing the customer's upcoming investment. This was a completely new approach in this sector. Primarily with the help of a net-present-value analysis – comparing life-cycle costs and life-cycle revenues – the anticipated financial consequences of their investment decisions were made clear to the customers. By using a scenario technique itself based on customer specifications, a range of potential market situations was simulated to highlight the potential return on investment: what happens if the order sizes over time become smaller, and what if the input material costs rise? With the modules created by Leistritz for providing holistic investment advice, technical evaluation of plant purchases is complemented with a presentation of the financial consequences – these being just as crucial, if not more so, to any company.

Example of a Customer Enquiry

To illustrate the process of offering such integrated economic advice, an example of an actual customer enquiry follows. The query addressed the number of plant changeovers it would take for the higher purchasing costs of the Flexxline extruder to be worth the investment. A set of calculation alternatives was presented to the customer. Initially, the cost of a standard extruder with its inherent changeover times was calculated. This was compared with the Flexxline system. The formulations for both plants were treated in the same way; in other words the sales revenue was defined only by a larger effective production time in the case of the Flexxline alternative. On the cost side, the changed variable input costs were taken into account.

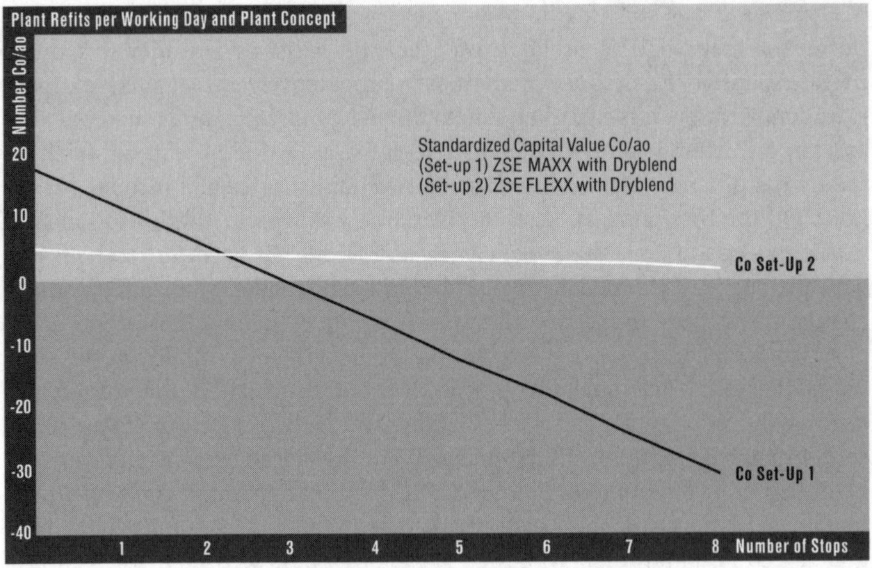

Fig. 3: Plant Refits per Working Day and Plant Concept

This information was transcribed as a series of payments across the entire service life of the two plants. As a final step, the different plant prices (the Flexxline is around 25 percent more expensive) were taken into account. Then the payment surpluses of the various periods were deducted from the face value according to the NPV method. Finally, the lot size and thus the number of required changeovers was varied. Fig. 3 illustrates that even at only about two plant stops per shift, the Flexxline is a better value when compared with the standard option. The example is based on reducing the changeover time by 80 percent. In the example, this amounts to reducing 2.5 hours to 0.5 hours. In practice, where a refit of all machinery is required, about 4 to 6 hours in total should be assumed, depending on the size of the overall plant.

The NPV calculated per machine for the given number of stops was related to the investment costs P in the interests of a better comparison and shown as a standardized figure [NPV_0/P] with no unit as the *y*-axis of the graphic. Based on this calculation, and after discussion of the customer's anticipated future order quantities, they overcame their initial skepticism and opted for the Flexxline product.

Interview with Richard Steiner[177]

Mario Rese

What has been your experience since introducing the Flexxline series with the new concept of selling systems in conjunction with providing financial advice?

Certainly a mixed bag. From the very beginning it was clear that there was enormous interest in new technology from lead users and early adopters in particular. These people take a really professional approach to addressing the issue of refitting times. In the past, discussions in our industry always centered on increasing the throughput rate of machinery. The extent to which productivity can be increased in these changed market conditions by more flexible use of machinery only became clear to us after an economic review of the whole customer process chain.

Did this different form of sales argument always meet with open ears from corporate customers?

Definitely not. And here it's not a matter of accusing players in this sector of being shortsighted. As long as all machinery providers were offering roughly the same refitting times, it was entirely appropriate to see throughput and price as the crucial parameters. It wasn't until technical alternatives emerged with a positive effect on refitting times that a rethinking of these considerations was required.

How did customers react to the fact that you were also providing them with advice on financial issues, of all things?

Admittedly they were initially quite skeptical. It took several years of work convincing them. We used financial calculations to compare the relative merits via the scenario technique in order to present these very technically minded decision makers with the overall financial consequences of their purchasing policies. It was a matter of slowly getting through to them by "hammering the message home," and we did this with a great deal of persistence.

Where did the work of convincing them actually take place? In customer sales talks?

No, that would definitely not have been sufficient. And as plant salesmen, absolute credibility during the actual sales talks can never be taken for granted. Initially we broached the subject at seminars and conferences to discuss the idea and of course gradually plant the seeds. But naturally we also learned something from the process, since weak points of the argument were revealed

Bringing Technology to Market. Olaf Plötner and Robert E. Spekman

ISBN 978-3-527-50270-7

during these discussions and showed us where there was room for improvement. Customer-specific workshops were then the second step in demonstrating the practical consequences of our modified technology to individual customers.

Where do you stand now in your efforts?

Scenario-based profitability analysis is now being discussed with a great deal of enthusiasm in our sector. It is clear that this planning and decision-making tool is becoming more and more readily accepted. The best sign is that customers are increasingly initiating this form of financial analysis themselves. And that gives us as plant salesmen the opportunity to take a much more multi-facetted approach to sale pitches–offering more than just technical arguments. The noticeably positive effect is a much greater flow of information on the part of the customer. This means it is not only the consulting service for the customer but also the chance for us to gain information that makes this approach so interesting. Helping a customer add value to the process chain is dependent on a very trusting customer-supplier relationship. Dealing discretely with confidential information is of course fundamental to this relationship. The hope is that customers will not be quite as open to all plant suppliers as they are to us.

How significant in your opinion is the consulting concept to the sales success of the Flexxline Extruder?

It's hard to overestimate. In purchasing new technologies, whether it be materials processing or plant engineering systems, customers like the security of putting all their faith in one supplier. Flexxline goes to the heart of the value chain of our customers. To reap the full benefits of this concept, their entire internal production logistics have to be changed. This means there is even more opportunity for our consulting services to reap rewards. With many new projects, Leistritz has been specifically asked to come back and present sample calculations at a separate meeting in order to demonstrate the financial consequences of converting existing plants. This is not only an additional opportunity to keep the communication channels with a customer open, but also a chance to meet other members of the purchasing team in person.

How does technical staff in particular react to your approach?

If your question is hinting at the skepticism engineers are reputed to have for business management considerations, I can only say: no problem. As we discovered, the NPV calculation is actually very appealing to technical staff, since it includes hard facts from areas like production and generates a finding that is quantifiable in euros and cents. However, and this was very apparent, technical people quite instinctively are not content with one set of figures. We have found there is much greater acceptance of the arguments if a range of scenarios is presented and the results illustrated in graphic form. It is the very act of openly addressing sensitivities that makes the evidence more compelling, to both technical and purchasing staff.

Were there any Leistritz in-house acceptance problems?

There are always a few people harboring concerns. Especially when a higher price is envisaged, many immediately see it as a threat to future sales and the imminent ruin of the company. This is particularly true in a market so heavily dominated by price considerations. For this reason, it was also important to find advocates within Leistritz for the new technological approach. The arguments were developed and honed during many in-house meetings, which was ideal preparation for later going public with the idea.

What further developments do you envisage? Is your concept easy for competitors to copy or have you built in some kind of copyright?

Copy protection is one thing. The much more crucial matter is what the Flexxline concept does for our branding image. The Leistritz company was the first in the market to address the topic of refitting times and flexible systems, which bolsters our position as the cutting-edge technology leader. Copy protection these days inevitably only lasts a short time; in the world of technology, there are always ways around these things. But in our market, long-term customer relations based on trust play a much more important role. Complexity may well be a barrier to sales, but it is also an entry barrier to our competitors. Any firm that wants to command respect as the cutting-edge technology leader cannot afford to rest on its laurels and must constantly think ahead. This makes Flexxline a unique kind of playing field (or think tank), where we can develop the compounder series of the future in consultation with our customers. It is this very combination of first-class engineering services and good business management sense that generates customer benefits. Happy engineering and happy selling do not work indefinitely–not in our market anyway, and probably not in B2B circles either.

4
The Services Shift in the IT Industries Case: IBM/SBS

Olaf Plötner

In the 1960s and 1970s, the IT market was dominated by hardware producers (HW), a business branch marked by relatively complex projects, most of which were ordered by public institutions and a few large organizations (e.g., NASA). Normally in those years, the necessary software (SW) was also produced by HW companies according to the requirements of individual clients. The resulting IT solution was sold as a package, including product-related services such as installation and maintenance.

In the 1980s, two technological developments caused major changes as far as those industry structures were concerned: the emergence of personal computers and the possibilities of client-server architecture. These developments resulted in a dramatic increase in IT technology users; in addition to generating private buyer groups, they above all revolutionized the use of IT at the workplace. Continuously new application areas were discovered where IT could support the daily tasks of employees in various functions. Even small and medium-sized companies were suddenly able to afford their own IT solutions.

With the increasing application areas, more and more SW products entered the market, while at the same time improved HW allowed for the introduction of increasingly substantial and useful software programs. In the course of this development, many new companies emerged (e.g., Microsoft). In contrast to companies from previous decades, these focused exclusively on the development of specific SW products. When it came to marketing, these suppliers wanted their products to be compatible with those of different hardware producers. In addition, clients insisted that different SW products be compatible with each other. Technologically, this became possible through an abandonment of proprietary solutions in favor of "open standards."

The possibilities of complex product integration, however, often exceeded a client's technological know-how, which led to suppliers designing, installing, and maintaining the entire architecture of their clients' system solution. The demand for these integration services was enormous. It grew much faster than the demand for HW, and in 2001 the turnover of IT services was higher than SW and higher than HW. HW and SW companies and consulting firms eagerly entered the market for integration services.

In the 1990s, numerous client companies experienced cost pressure; very often concentration on core competencies became their strategic motto. This was accompanied by a desire to no longer provide personnel for the increasingly complex IT solutions. In order to manage with variable costs, companies

Bringing Technology to Market. Olaf Plötner and Robert E. Spekman

ISBN 978-3-527-50270-7

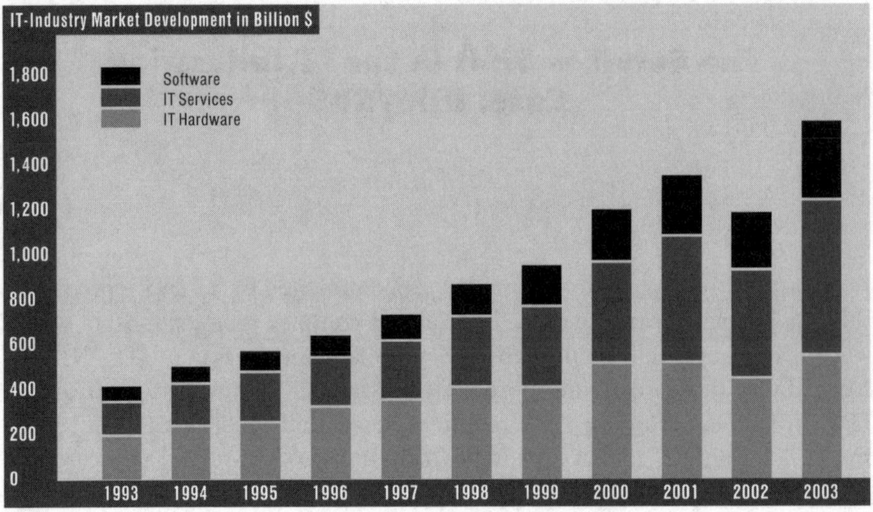

Fig. 1: IT Industry Market Development[178]
Source: EITO and IDC

started to look for specialists to handle these IT activities more efficiently. In this period, many client companies followed the example of the few that had already decided to outsource their IT departments. This meant largely dissolving internal IT departments and offering their workload to external suppliers.

This trend was accompanied by a decrease in the purchase of IT equipment by the client companies, which by then were more interested in buying certain IT processes. In these cases, suppliers often re-employed the IT personnel of a client company, even though – given the advantages of their efficiency – they did not need the same number of employees. Rather they expected to employ their newly acquired workforce for a different outsourcing client – provided, of course, they were able and willing to be so employed. The suppliers of outsourcing services were often the same companies that sold overall system solutions. There were, however, many (such as EDS) that had pursued this kind of business model since their establishment.

At the end of the 1990s, new application areas for IT solutions continued to be developed. In particular, the broad acceptance of the Internet and the increasing potential of e-business provided further possibilities for company use. Accordingly, the outsourcing business grew considerably.

By then IT outsourcing was often no longer limited to IT processes as such. With increasing frequency, administrative activities such as factoring and sending monthly salary slips or client invoices were also being outsourced. More and more suppliers were being asked to handle production processes that were highly IT-related processes, such as account transactions in banking. Payment by the outsourcing supplier frequently depended on the market success of the products sold by the client. Companies, for example, that had outsourced the handling of payment transactions to a bank paid according to the number of

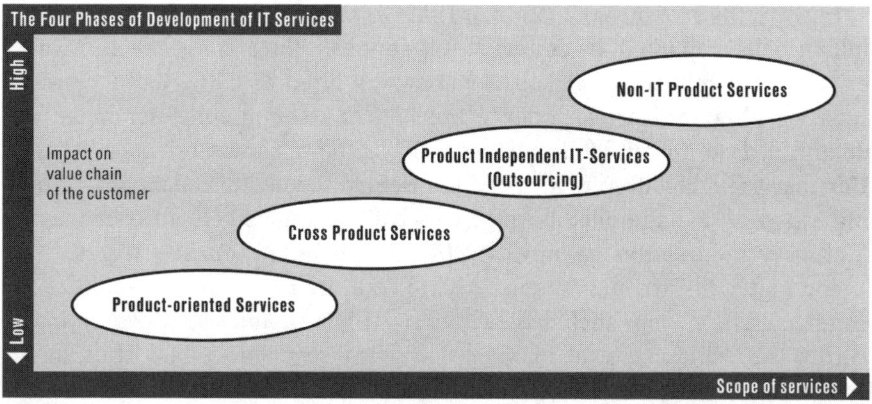

Fig. 2: The Four Phases of Development of IT Services

transactions. This meant that if the bank prospered and acquired more customers making a lot of transactions, the outsourcing partner prospered as well. (The same process of course took place inversely.)

IBM Global Services

First appearing in 1924 under the name International Business Machines Corporation, IBM soon turned into a major, internationally renowned company, responsible for the development of many patents leaving a permanent stamp on IT technology, along with discoveries sometimes leading to a Nobel Prize.[179] In the 1960s and 1970s, IBM was the biggest IT company in the world.

As we know, however, IBM was not able to participate in the enormous boom the IT industry enjoyed in the 1980s. Although there was still no other company boasting such comprehensive IT competency, IBM's economic situation increasingly worsened at the end of the 1980s and beginning of the 1990s. In 1991 the company was still the market leader, with a turnover of 64.8 billion dollars, about 9 percent of which fell to the marketing of IT services. Nevertheless, the turnover had decreased compared to the previous year, with a net loss of 2,861 million dollars.

Since the growing market of IT services looked promising, the managing board of IBM decided in 1991 to strengthen this business line, announcing a "new worldwide services strategy." The explicit goal was to turn IBM into a world-class services company by 1994. At the same time, IBM restructured its Systems Services Division into the Integrated Systems Solutions Corporation. The expectation was that the large number of installed IBM solutions worldwide would mean a large advantage in successfully installing this services business.

In 1993, the economic situation of IBM worsened dramatically. The loss of 8 billion dollars marked an annual result seriously threatening the company's existence. In this situation Lou Gerstner was hired as CEO. Gerstner was a man with low industry experience but high marketing competency; he was determined to forcefully implement IBM's new, service-oriented strategy. Gerstner brought with him a customer-oriented sensibility and strategic thinking expertise he had honed during years both as a management consultant at McKinsey and as a top executive at RJR Nabisco and American Express.

Soon after his arrival, Gerstner had to take dramatic steps to stabilize the company. These steps included rebuilding IBM's product line, continuing to shrink the workforce, and making significant cost reductions. Thus in the course of the 1993-94 re-engineering processes, 36,000 employees were dismissed, equaling 14 percent of the workforce. On the other hand the company put additional funding into its consulting group, which was unveiled in 1992. This build-up reflected a shift in IBM's services philosophy from focusing merely on after-sales activity such as maintenance to a broader stress on assisting customers in every facet of their ongoing business operations. Parallel to the restructuring of human resources, expenditure for training courses and executive education was substantially raised.

Despite mounting pressure to split IBM into separate, independent companies, Gerstner decided to keep the company together. The 2001 annual report of IBM put it as follows: "We were on a fast track [in the early 1990s] to being dismantled from within." "We believed niche players weren't the future," the report continued, then explaining that "in fact, breaking up the company would have been the end of everything IBM stood for. We made a big bet that customers needed a partner who could both create technologies and integrate them – with each other, and with the customer's business processes. At the time it was a gutsy call. They always are when you're alone. But we decided that we should be true to ourselves. It all started with that."[180]

The 1993 decision not to break up the company became the basis of an "integrator" business model for IBM. IBM Global Services has become the focal point of this model, bringing together hardware, software, and services. Gerstner decided that IBM would be vendor-neutral and recommend any products of rivals like Microsoft, HP, Sun, and other major competitors if they happened to be the best solution for the customer.

A second strategy devised in the early 1990s built on the client/server phenomenon of the late 1980s and predated the explosion of the Internet for commercial use. IBM initially called it „network-centric computing". In 1994, harnessing the company's substantial networking capabilities, IBM brought together multiple internal networks under a single organization to form the IBM Global Network. Its mainstay was electronic data interchange services.

In 1995, IBM formed a new unified organization, IBM Global Services (IGS), to provide customers with worldwide reach and more consistent service levels. To this end, Global Services rationalized its offerings, moving from an ungainly 2,500 offerings worldwide to about 100 solution categories that were consistent globally. In December 1996, Dennie M. Welsh, the first general manager of IGS, commented on this development as follows: "It is clear we cannot stand still. IBM Global Services must change because our customers, marketplace, and competitors are all changing. Our new management structure will enable greater teamwork throughout IBM and continue building on the momentum of profitable services growth and customer satisfaction worldwide."[181]

Among the changes that then took place were reorganizing services around clear business lines mapped to customer needs and concerns; aligning IGS capabilities, skills, and practices around three major areas (innovation and integration, infrastructure support, and strategic outsourcing); and gearing up consulting group practices to help clients implement and manage standards-based environments.

IGS was growing customer by customer and investing to hire and train experts in everything from IT consulting to systems architecture and web services. IBM used its financial strength to fund its expensive push into outsourcing, and the company placed informed bets on the future in areas such as IT utility services (e-business on demand) and hosted storage. In fact, IBM had already started its outsourcing business in 1989, when it not only designed and installed an IT solution for Eastman Kodak in the United States but for the first time took responsibility for running such a system. The success of this project encouraged those in charge to provide additional outsourcing services by exploiting the excess capacity in their own data centers.

IGS made other adjustments as well. It moved out of the network infrastructure business by selling Global Network because it could acquire underlying connectivity from other providers at less cost and investment. As a corollary, IBM chose to concentrate on managing the information and data that was streaming into networks by connecting business processes across enterprises and up and down industry value chains. In addition, IBM merged its consulting group – which had been tied to its sales units – into Global Services. Ginni Rometty, general manager of IBM Global Services, Americas, has observed that "what this did for services and for IBM was to join at the hip our best market-facing thinkers with our integration experts who knew how technology fits together to deliver higher value solutions to the customer."[182]

At the start of the new century, IBM Global Services expanded its portfolio to offer many more e-business services, in a reflection of the powerful effect of the Internet on commerce. Customers needed help in sorting out the new range of technology choices. They wanted advice on a coherent strategy for e-business and a better understanding of the impact of technology on their overall enterprise strategy, especially advice relevant and specific to their own industries. Much of what IBM learned and applied to customers was derived from the company's own transformation during the late 1990s into a leading e-business in its own right. IBM converted its in-house experience in e-business into several commercial opportunities. For example, internal advances based on business intelligence and knowledge-management tools were incorporated into many solutions; and IBM's expertise in using an intranet as its primary internal communications channel paid dividends when many customers asked the company to build similar web environments for themselves.

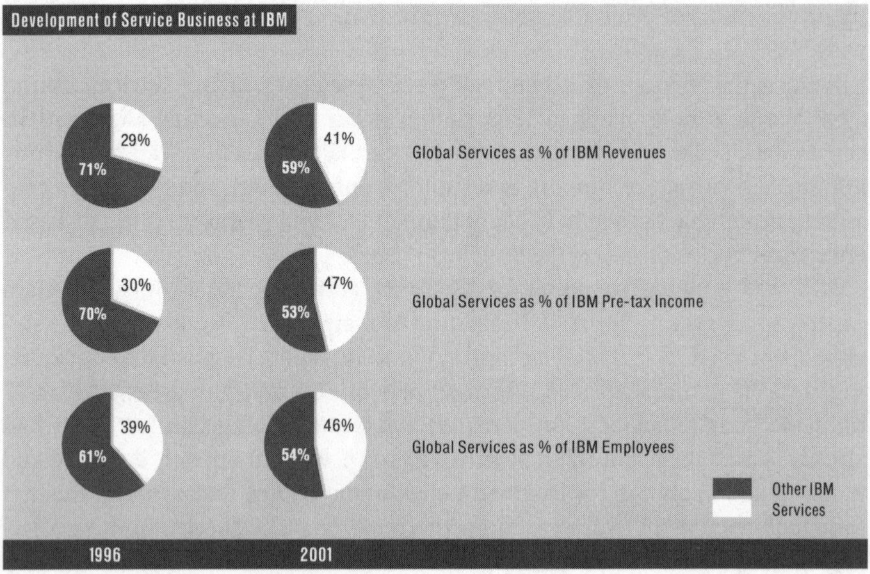

Fig. 3: Development of Service Business at IBM[183]
Source: IBM Annual Reports 1997 and 2002

In 2001, the business of IT services alone generated more than 40 percent of IBM's 86 billion dollars in sales. That business had thus emerged as the single largest source of revenue in IBM's portfolio (see figs. 3 and 4). But despite this heavy increase in the service business, IBM continues to own a strong R&D unit, into which the company annually invests more than 5 billion dollars. The general orientation of this unit, however, has changed considerably. In the late 1990s, for example, management began to encourage its researchers to make regular visits to customers–in some cases to help solve a particular problem; in other cases simply to better understand how enterprises actually use information technology.[184] Research efforts were oriented to help solve problems in areas such as business design transformation, process integration, advanced supply chain management, and risk analysis.

Samuel Palmisano, who after his successful years at IGS was made IBM's new CEO in 2002, has explained that "IBM Research has reinvented itself many times over the decade as technical challenges and customers' problems have changed. A generation ago, most of our researchers grappled with challenges in physics, materials science, and magnetics because so much of our business depended on devices. ... They understand that the business problems we are trying to solve for customers are not elementary stuff–it's hard, hard, hard. ... Technical leaders want to make an impact, which means seeing that their work makes a difference in the real world. Sure, they thrive on publishing and going to conferences. But they understand that there's no better way to do that than working with real, live customers."[185]

Going forward, IBM Global Services has increased its industry focus by concentrating its skills, knowledge, and capabilities on clients' specific business issues and building an inventory of reusable assets to speed response time. IGS has integrated at the industry level, where it concentrates on sixteen industries grouped into five sectors. In 2002, to strengthen its problem-solving capabilities, IBM announced the purchase of PriceWaterhouseCoopers Consulting (PWCC) for 3.9 billion dollars, signaling the biggest acquisition to advise customers on business strategy, not just run their IT systems. It combined 25,000 IT-focused consultants in IBM's Global Services arm with a roughly equal number from PWCC (there were 3,500 layoffs) and created IBM Business Consulting Services, a 13-billion-dollar business comprising almost a third of the overall services group. PWCC was organized around the industries it served.

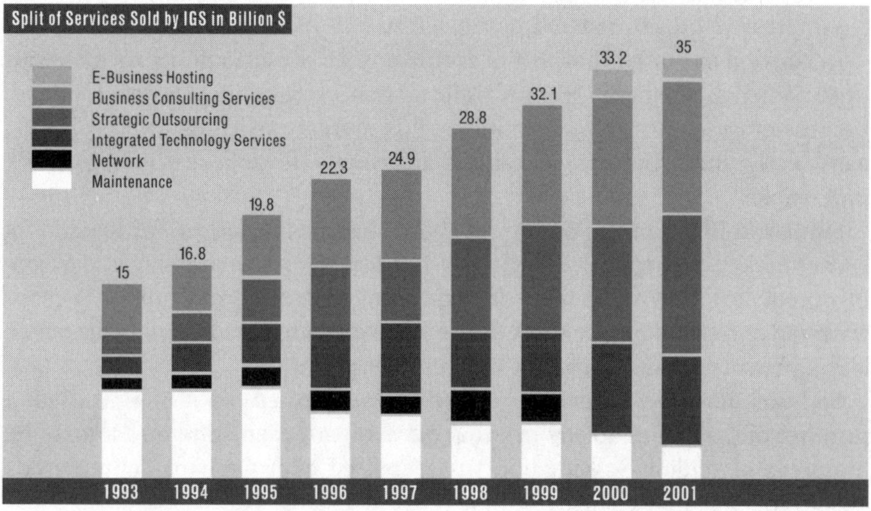

Fig. 4: Split of Services Sold by IGS[186]
Source: IBM Annual Report 2005

Siemens Business Services

Siemens Business Services GmbH (SBS) was established in 1995, even though its roots go back to the 1990 merger of Nixdorf AG and Siemens Business Unit Data Systems.

Nixdorf AG had been founded by Heinz Nixdorf in 1952. Until the mid-1980s he was in charge of an organization that had become the fourth largest computer company in Europe. Like its charismatic owner, the company was thought of as possessing first-rate engineering acumen and being innovative and creative. Nixdorf produced a large assortment of hardware and software and marketed such products together with related services. These marketing activities were mainly in Germany but also were in several other Western European countries. Large clients consisted above all of banks and business enterprises with a widespread system of branches; these enterprises bought comprehensive IT systems to optimize their business processes. In this manner Nixdorf AG acquired, for example, an excellent reputation for installing nationwide grids for servicing automated tellers.

Siemens Data Systems had been one of the seventeen business units of Siemens AG. During its 150-year-old history, the company had turned into an internationally renowned organization, which was due–together with its data systems–to its technical products and its success in areas including telecommunication, energy, transport, medical equipment, and automation. With a turnover of 40 billion dollars, Siemens belonged to the fifty largest companies worldwide as early as 1990. Compared to the other units, the turnover of Data Systems was relatively small, but it had clearly grown at a faster rate. However, the impressive growth rate did not match its rate of profitability.

In 1990, hand in hand with the acquisition of Nixdorf and its merger with Data Systems, Siemens Nixdorf Informationssysteme (SNI) was founded. Because of different corporate cultures, this merger carried its own problems, particularly since German legislation made workforce restructuring nearly impossible.

Similar to Bull (France) or Olivetti (Italy), SNI had its largest turnover in its home market. SNI regarded itself as a full-scale computer producer, developing, producing, and marketing hardware along with basic and application software and corresponding services. Since SNI worried about losing its profitable leasing business, outsource projects were not permitted.

SNI was, however, never able to reach its anticipated profit margin. With a turnover of 6.2 billion dollars in 1994, the company even suffered a loss in the hundreds of millions–a situation that persisted after Gerhard Schulmeyer's appointment as new chairman of the board that same year.

In 1995, the Siemens central board agreed to establish Siemens Business Services GmbH (SBS) as a daughter of SNI AG and Siemens AG. The company was meant to advise clients on IT matters, in other words the design, installation, and operation of the necessary system solutions. Friedrich Fröschl was appointed president and CEO. Fröschl was a 43-year-old executive who came from CSC, the American IT services company, and had not worked for Siemens previously. That same year, Siemens turned its German main computer center into the business responsibility of SBS. Likewise, all departments in SNI that were not clearly focused on hardware or software were transferred to SBS. In the process, less profitable units were shed and the company's portfolio consolidated. The turnover of SBS in its first year amounted to 628 million dollars. Ninety percent of this turnover was due to purchases by Siemens AG.

Other than was the case for SNI, outsourcing projects were explicitly open for SBS. Several IT organizations were thus acquired from companies purchased by proprietary systems of SNI, the customer group thereby being limited to Germany and several western European countries. Following the outsourcing approach, either the responsibility for all internal Siemens IT organizations was assigned to SBS or the employees reported to SBS.

In 1997, the solutions business was transferred from SNI to SBS, which meant that the old SNI was more or less dissolved. The challenges of this merger were comparable to those of the 1990 merger, particularly since not only were all SNI central departments integrated into SBS, but software application as well. In 1998, the turnover of SBS increased to 3.5 billion dollars. But in the same year losses had to be reported for the first time, after initial years that had been profitable. Based on its turnover, SBS had climbed to become the ninth largest provider of IT services worldwide. Table 1 shows the 1998 provider structure.

Leading IT Services Companies and Their Market Share, 1998

World Market IT-Services 1998

Position	Market Share	Position	Market Share
1. IBM	8.0%	6. Fujitsu	1.7%
2. EDS	6.0%	7. ADP	1.6%
3. Anderson	3.0%	8. Cap Cemini	1.5%
4. CSC	2.5%	9. SBS	1.2%
5. First Data	1.7%	10. HP	1.1%

Table 1: Leading IT Services Companies and Their Market Share, 1998[187]
Source: Analysts' Meeting 1999

Finally in 1999, the Technical Services SNI unit was transferred to SBS. The continuously strong growth of SBS, however, was basically due to a strong increase in outsourced business. For example, SBS won a 1.5-billion-dollar outsourcing contract with the National Savings Bank in England, running for 15 years. In 1998-99, this was the third largest outsourcing project in the world. (The biggest was a 3-billion-dollar contract between IBM and Cable&Wireless.)

To enhance the growth rate of SBS, the synergy potential of SBS and other Siemens units had to be exploited. Very good opportunities were seen in the highly successful telecommunication grids, where IT had by then acquired paramount importance. The same was true for power plants (where the client demand for complex invoicing services increased) and transport (where there was an increased demand for IT-based ticketing systems). New business for SBS, in other words, was meant to be generated through the high expertise and first-rate customer relations of the various Siemens units. The Siemens Central Board explicitly supported this development, but a comprehensive incentive system for cross-departmental cooperation was not installed.

In 1999 and 2000, the SBS turnover continued to grow. Profit was made, but the contribution margin did not meet the corporate target. The year 2001 produced losses, which were above all due to high provisions tied to the National Savings Bank project. After six years as managing director of SBS, Friedrich Fröschl was replaced in October 2001. At that time, SBS had a turnover of 6.8 billion dollars, 33,000 employees, and branches in more than sixty countries. Fifty percent of the company's turnover was made in Germany, 25 percent of this by Siemens. Whereas Siemens units like Power Generation or Medical Systems clearly made it back into the profit range after a difficult time, SBS would remain a problem of the Siemens Central Board over the following years.

The author is grateful for the valuable input offered by Friedrich Fröschl.

Interview with Friedrich Fröschl[188]

Olaf Plötner

Mr. Fröschl, looking back what is your view of the success of SBS and IGS between 1995 and 2003?

IGS was and remains very successful. It is the most important benchmark for many companies in the industry. From an economic perspective SBS saw a convincing growth in sales, but with regard to profits the central board's expectations were higher than what was actually earned.

What for you was the most important difference between the starts of SBS and IGS?

A decisive difference in the opening phase was certainly that for IBM the success of IGS was directly tied to survival of the entire firm. In contrast, for Siemens AG SBS was always only one of many parts of the concern, meant to fit into its overall strategy and not endanger the big picture. When it comes to taking tough measures with restructuring personnel or a readiness to take financial risks, that plays an important role.

Surely a willingness to take financial risks is needed in the first phase of operating complex IT systems.

That's correct. Especially when – as was the case with us – employees have relatively little experience in the business. Until 1995 outsourcing with clients in the IT branch was even strictly forbidden – the belief was that the product business could be protected in that way.

In this respect, higher risk-taking willingness was not the only advantage IBM had over SBS. What was important above all was the fact that through its globally strong HW business, the company had a much larger installed base available worldwide. Its product know-how gave it advantages in building an international network of customers.

As a customer SBS could in any case use the other divisions of Siemens AG from the start. It more or less had a purchasing obligation. Wasn't that an advantage?

With regard to the first sales numbers, definitely. But together with the purchasing obligation come claims and concerns that can be completely counterproductive. Internal customers are often more difficult than external ones, and it's interesting to see that the IT service firms that emerged from large industrial firms were less successful when they had high in-house participation.

Bringing Technology to Market. Olaf Plötner and Robert E. Spekman

ISBN 978-3-527-50270-7

Did this also contribute to SBS's inability to realize its synergies as hoped for in view of the worldwide customers of the other Siemens divisions?

Yes, high internal acceptance is a prerequisite for the development of external customer relations, especially when colleagues are supposed to make *their* customers available to another division.

Did the fact that a newly established division within a large firm first having to struggle for internal acceptance also have something to do with SBS having to simply take over all of Siemens Nixdorf's employees?

There are a number of other reasons for this – they are mainly tied to factors involving Germany's legal framework and an understanding of Siemens AG's social responsibility. But these points should be viewed by no means only in a negative light.

In any case, in its decision to force through changes in its service business, IBM dismissed 36,000 employees of the workforce active in the product business. Doesn't a change of business focus have to be linked to a change of employees?

Certainly not a change of all the employees – even IBM didn't do that. For IGS, the product know-how of the employees was an important advantage in the development of the service business. But alongside product know-how there's a need for other sorts of expertise that are not always present in product-oriented firms to the same degree.

What sorts of expertise do you mean here?

Above all there's a need for extensive process know-how. This must then be applied in a structured way according to customers' needs.

With regard to SBS's IT business, I distinguish four realms: designing, building, operating, and maintenance. The need for maintenance means providers have to offer superior logistics along with relatively standardized processes that provide a framework in which the employees work through their tasks. Operating mainly involves adapting to unexpected difficulties and promptly adopting the correct measures. Flexible employees who can grasp things quickly and are strongly committed are needed for handling such exceptions. On the other hand, building IT solutions for the customer calls for good analysts who have a sure sense of technical potentials and the possibilities of various process architectures. Finally, designing customer solutions works best with advisor types who understand customers' multi-leveled problems and can communicate in a suitable way that generates trust.

Within these four groups, provider-firms have to define sensible modules on the basis of available know-how, so that complexity and thus costs are kept under control. Beyond this the four groups are linked through a need for permanent customer orientation in thinking and acting. And precisely here, employees coming from a more or less product-oriented world often have difficulties.

In your view how can the thinking and acting of these employees be changed?

It's important that from the start the new demands are clearly communicated to the employees. Then they need to be asked if they'd like to develop in that direction, and you have to analyze the individual developmental needs, responding to the latter with concrete offers for further development. In this industry the process of continual learning and development takes on surprising importance. In the long term, an employer who doesn't offer the necessary possibilities in that respect has no prospects of success.

Can these developments be successfully dealt with by all employees whose employers are shifting their firms from product orientation to service orientation?

No, it's certainly the case that not everyone can change to the required degree. Leaving SBS aside, against the background of my experience with numerous firms I can say that with around a third of the employees such a developmental process takes place virtually by itself; with another third by managing it with support. The final third either cannot cope with the new demands in a lasting way or doesn't want to – which is entirely legitimate.

Looking back on your activities as a CEO at SBS, what have been the most important lessons learned from that period?

For one thing I'd certainly manage the expectations of the shareholders in a different way. As an aside I'll mention that I'd here make a rule of consistently documenting all commitments and agreements. At the same time I'd certainly steer things more closely and crack down more firmly – that includes more carefully controlling whether employees have realized what's been planned. Although that's not to my taste, it's still something I've come to see as extremely important.

What has changed in the IT industry since 2003?

Above all globalization has advanced even further. The Indian IT firm ITS has now outsourced a number of its activities to Romania. No one could have imagined that a few years ago.

The increased focus of many firms on their core competencies is especially important for the IT branch. Against that backdrop, for a long time now they've been letting outside companies take care of not only IT-related activities but others as well. They here use the term *business process outsourcing*, meaning something like moving the entire accounting or purchasing division or HR administration elsewhere. Even auditing firms and investment banks now have many of their analysis-based activities taken care of by firms in India. Deutsche Bank is only one example of this among many. Since a large number of these outsourced processes are dominated by IT, many new opportunities have emerged in this way for IT firms. I should mention that IGS has been using these opportunities in a very professional manner.

And what developments do you see in the industry over the next five years?

First of all I believe that the past years' developments – the increase in both globalization and focusing on core competencies – will advance even further.

Viewed technically, the long-predicted convergence will now become reality. IT, telecommunications, media, and other formerly separated industries will merge. In the process many new developments in technical application will become part of daily life.

On the provider side, rising price pressure will force service-oriented firms to work further on standardizing their processes. Through higher standardization, the possibilities will grow to individually fit the scalability of IT solutions to the needs of customers. Increasingly customers are receiving what is most sensible for them, in an efficiency-conscious framework.

With the growing technical and global possibilities on the one hand and the intensifying price pressure on the other, it is becoming increasingly important for provider-firms to keep costs under precise control. This will have a number of consequences; for example, the role of risk management in such firms will increasingly become more important.

5
Advanced Sales Management for a Complex Industrial Portfolio
Case: Siemens A&D

Aubert Martin
Dale Wilson

Siemens Automation & Drives (A&D) provides a comprehensive portfolio of products and systems for industrial automation and control, as well as for (low-voltage) power distribution for industrial, commercial, and residential construction. The portfolio comprises products such as motors, drives, controls, breakers, switches, programmable logic controllers (PLC), human machine interface, and process instrumentation.

The market for such a portfolio is highly diverse and covers all types of industries, including the automotive, aerospace, chemical, pharmaceutical, semi-conductor, food and beverage, metal and mining, and oil and gas industries. On the construction side, the market ranges from very complex infrastructures (for example, stadiums, airports, and hospitals) to multi-family and single-family residential buildings.

Siemens A&D is the worldwide leader in its business. Based on a very strong competitive position in Europe, it is aggressively pursuing profitable growth in

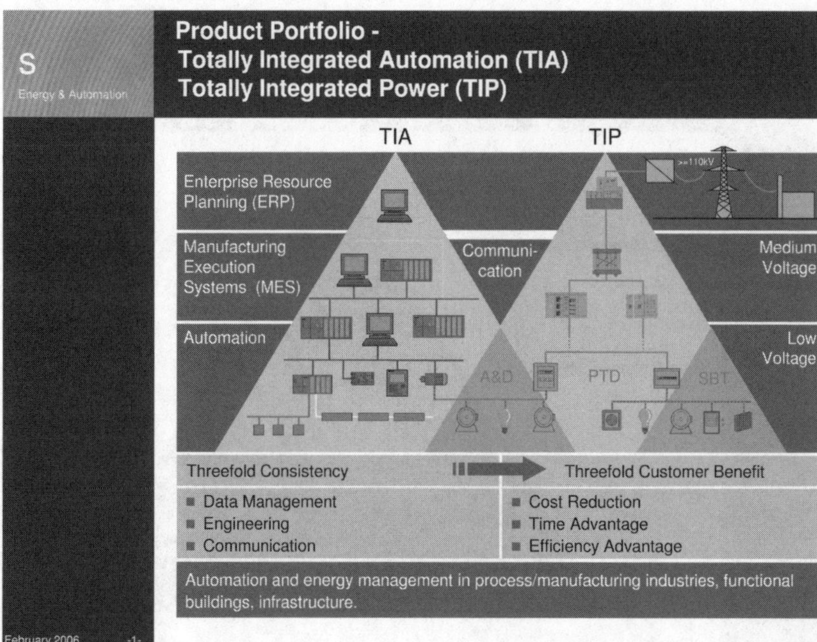

Fig. 1: A & D Product Portfolio

Bringing Technology to Market. Olaf Plötner and Robert E. Spekman

ISBN 978-3-527-50270-7

the other major economic regions of the world. Each regional Siemens company is responsible for executing the global strategy aligned to its local requirements through a customer-focused sales and support structure.

Siemens Energy & Automation (SE&A), headquartered in Alpharetta, Georgia, is a U.S. operating company within the global Siemens structure. It is responsible for the U.S. business of two Siemens groups: Automation & Drives (A&D) and Industrial Solutions & Services (I&S). In this chapter we consider only the A&D part of SE&A business, which is about 80 percent of the total sales volume. The U.S. market for A&D is approximately 20 billion dollars. Fig. 1 shows the overall A&D value proposition of Totally Integrated Automation (TIA) and Totally Integrated Power (TIP) – a comprehensive product portfolio, consistently designed for quantified and measurable customer benefit.

Selling such a wide range of products and services in a highly advanced and complex industrial market, against strong domestic and international competition, requires sophisticated sales management methods together with an appropriate structure and effective processes. In the following discussion, we present an advanced sales management model based on the Siemens A&D business in the United States.

Strategic Direction and Target Setting

The strategic goal of A&D is to increase sales at twice the rate of market. This goal is broken down into individual targets for product and market segments.

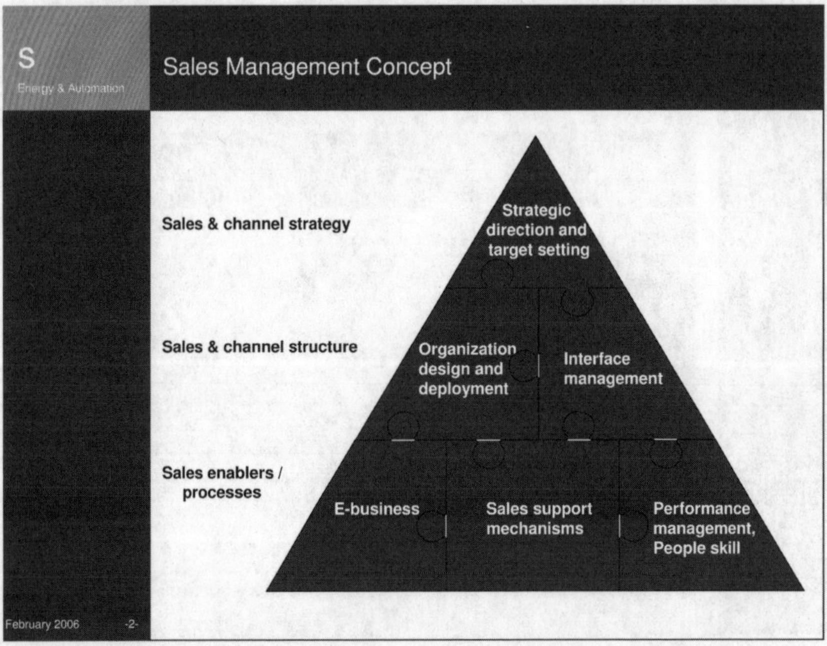

Fig. 2: Sales Management Concept

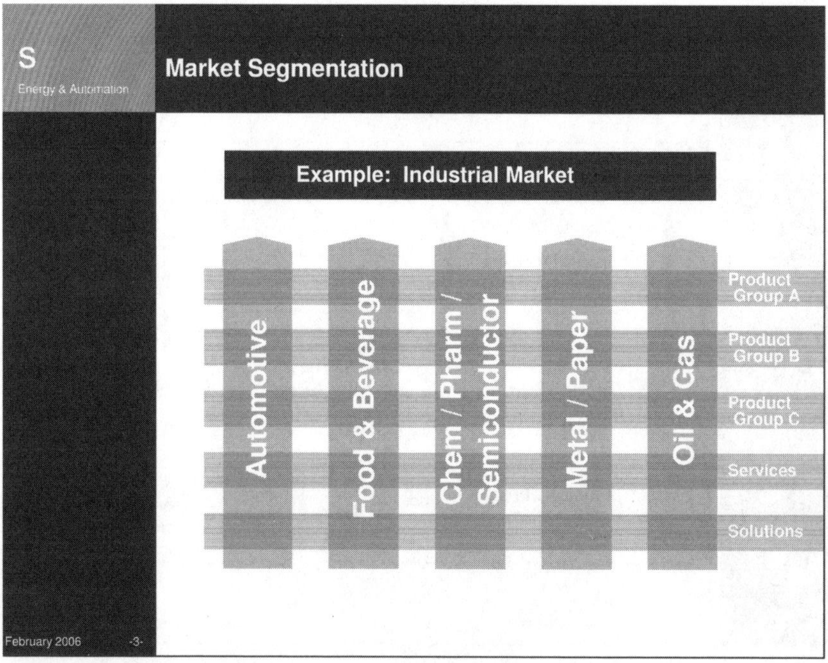

Fig. 3: Market Segmentation

The major business drivers to meet these goals are clearly defined and the appropriate metrics specified. Finally, this is documented in a balanced scorecard – a "roadmap for profitable growth" – for the company overall and for each division and business unit together with all other functions.

The company's sales management concept follows the same principle. Fig. 2 presents the major elements of the sales and channel strategy, the structure, and the sales enablers and processes. Initially, the strategic targets and directions have to be agreed on, based on the company's overall goals. In the strategic marketing process, market segmentation, value propositions, the channel of choice, and specific top-down targets are developed. Joint bottom-up planning between sales and marketing is used to evaluate the opportunities per customer in each market segment. Market segmentation is done by product group, industry segment, and region (see fig. 3). This is the groundwork for a consistent business planning process that ultimately results in business target agreements between the global and regional units and between sales and marketing.

Organizational Design and Deployment

Organizational design and interface management is critical for a successful go-to-market strategy. On the customer side, we differentiate between the construction market and the industrial market (see fig. 4), with customers spread widely across the continent.

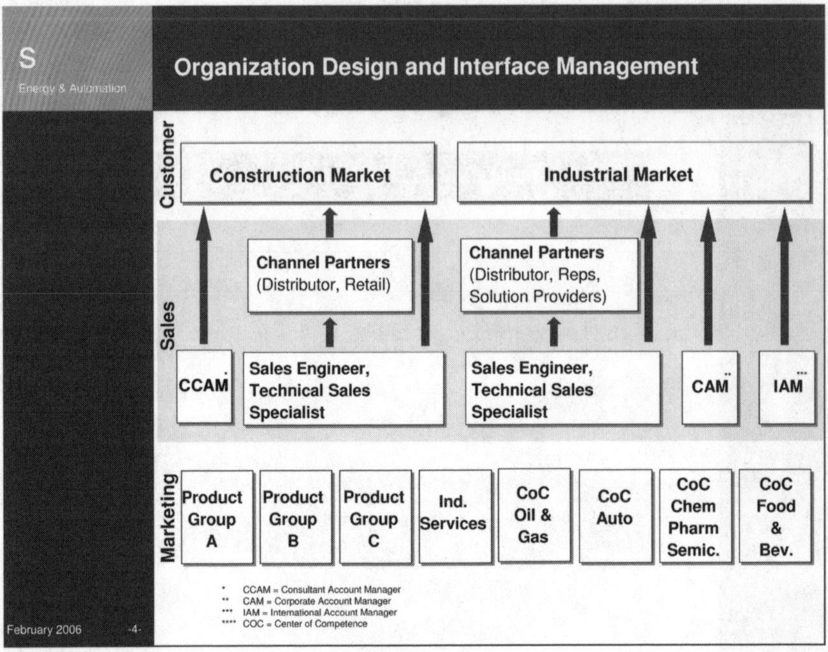

Fig. 4: Organization Design and Interface Management

To address this market effectively with a large cache of products and solutions, a multi-channel approach is necessary. Channel partners such as distributors, retail outlets, manufacturer representatives, and solution providers serve as multipliers of the sales force. They provide local services (for example, warehousing, logistics) and sometimes complete system integration. The sales force comprises account development experts (sales engineers) and technical system sales personnel (technical sales specialists); for the various products and solutions, we have established consultant account managers (CAMs) and, for large accounts, various levels of national and international account managers.

This organization is supported by product marketing groups and by centers of competence (CCs) for vertical market segments (for example, the automotive industry and food and beverage industry). The CCs have industry experts and develop industry-specific value propositions. They also champion the sales effort with large global accounts which have corporate account managers.

The direct selling organization is expected to create demand in the market for the products, solutions, and services that SE&A offers. To effectively use resources, customers are segmented by type and complexity (fig. 5).

Corporate, international, and national accounts require a high level of coordination due to the number of locations. These accounts are selected based on industry alignment and on potential revenue for SE&A. Industry-segment accounts require a high level of technical support, and they receive special attention in the form of application and solution support coordinated with the

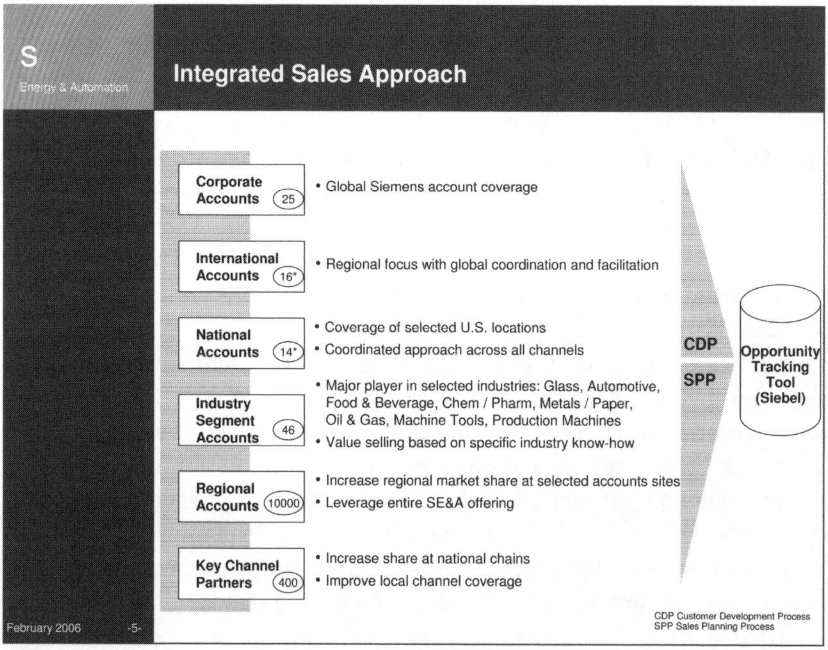

Fig. 5: Integrated Sales Approach

relevant center of competence. One of the largest challenges is gaining direct access to the thousands of regional customers and promoting a significant portfolio of products. A sales model that combines sales engineer and technical specialists allows the company to focus its attention on selected regional customers. A bottom-up sales-planning process (SPP) is used to identify and assign approximately 25 customers to each sales engineer. The goal is to significantly increase account share.

A customer development process (CDP) is a more in-depth planning process for selected accounts; it is used to qualify the customer opportunity and to develop a plan to maximize account share. The tracking of these opportunities is done using a customer relationship manager (CRM) tool (see below).

Channel partners such as electrical distributors and solution providers also play a key role in expanding our capacity for broader account coverage. To leverage the capabilities of these partners, collaborative business planning is required. Identifying business opportunities, setting joint target levels, and coordinating selling efforts is critical when building a complementary channel. SE&A has approximately 2,000 partners. However, planning efforts focus on a smaller number (400). This subset of partners contributes most to SE&A's overall performance.

Interface Management

Coverage of Large Accounts

Large accounts are typically national, international, or corporate multi-site operations with various functional departments (such as engineering, purchasing, manufacturing) that are decentralized throughout the organization. Buying influencers and decision makers are often widely dispersed both geographically and vertically within business units of such accounts, which rely heavily on a complex network of suppliers, service organizations, and supply-chain logistics providers to operate effectively. Sales management for the accounts requires coordinated matrix coverage (see fig. 6).

An automotive company is one example of a large account using multiple Siemens sales resources for collaboration on meeting common sales objectives. Corporate account managers provide C-level coverage and strategic positioning within the account, advance notice of capital expenditures for major projects, identification of preferred original equipment manufacturers (OEMs), and assessment of project scope to align other Siemens resources. National account managers and sales engineers are often assigned to remote sites of the account to engage appropriate buying influencers and decision makers, individuals who have an impact on projects, specification, and product application. Siemens' authorized channel partners and solution providers typically com-

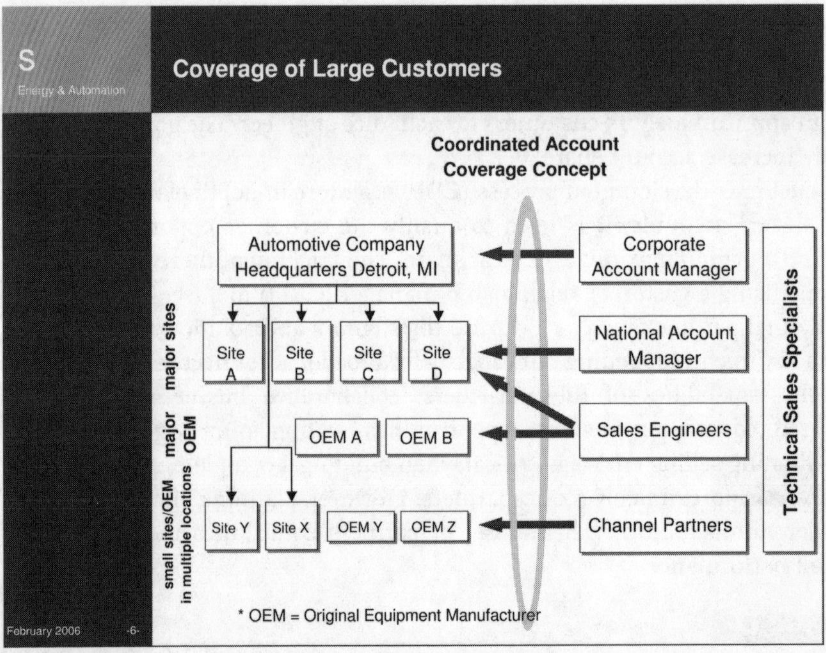

Fig. 6: Coverage of Large Customers

plete the firm's coverage of large accounts by selling and supporting smaller sites and secondary OEMs. Technical specialists work across the sales matrix as needed to address technical sales and support requirements.

Channel Partner Coverage and Segmentation

In order to achieve optimal performance from its channel partners, SE&A segments both their sales and service coverage (see fig. 7) and each partner's value to Siemens in specific market segments (for example, automation, industrial, motors, commercial and industrial construction, and residential construction) using the trademarked Market Builder Channel Partner Program.

As shown in fig. 7, a three-tier sales coverage strategy is used. It deploys national account managers assigned to develop C-level relationships at the headquarters of named global, national, and super-regional electrical distributors. Sales division regional managers, regional sales managers, and distributor specialists are focused on the top four hundred distributor branch operations across the United States.

By contrast, a centralized distributor-development group is assigned to contact and develop sales through the Siemens network of low-sales-volume distributors. This three-tier sales coverage ensures that the company's full range of channel partners receives focused and appropriate sales and service support,

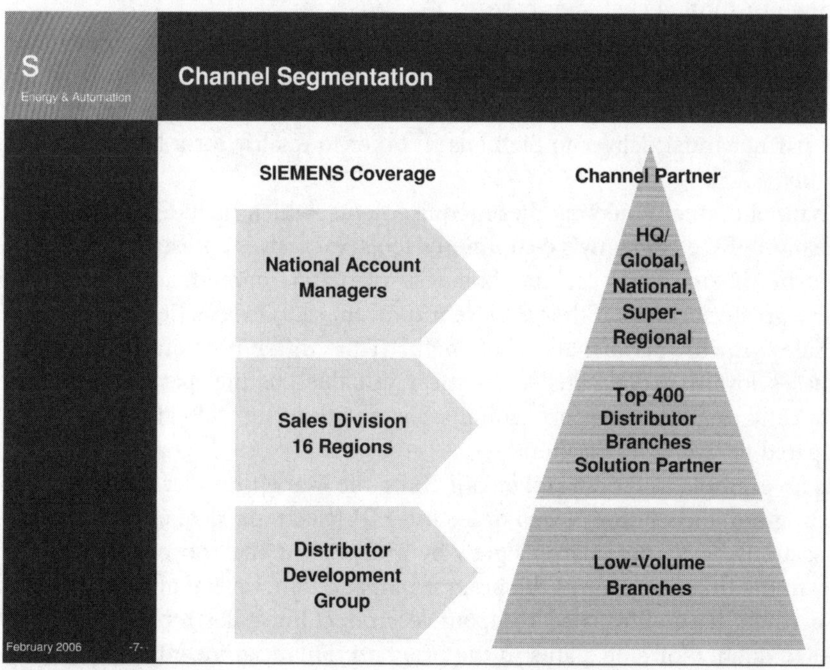

Fig. 7: Channel Segmentation

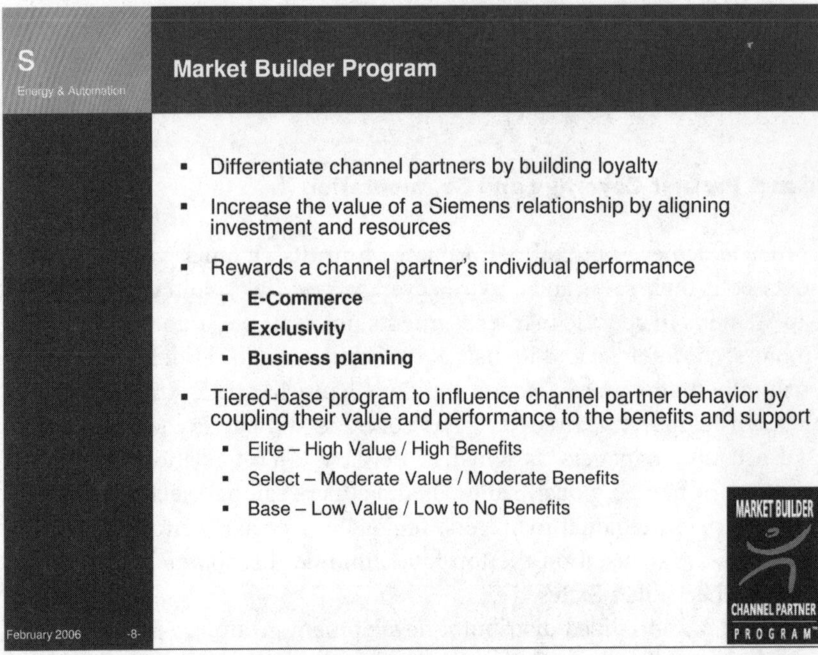

Fig. 8: Market Builder Program

all in an effort to help increase the partners' effectiveness in selling to and supporting our mutual customer base.

The Market Builder Channel Partner Program (see fig. 8) is a tiered value program that creates differentiation between the company's channel partners in each market segment; it does so by defining "value-add" requirements that each partner must deliver to Siemens in order to qualify for a higher level of benefits.

Channel partners meeting these requirements, which include adoption and utilization of the company's e-commerce tools, exclusive representation of Siemens in the marketplace, and business plan development and execution, receive greater benefits if they achieve minimum sales expectations, incremental sales-growth performance, or both. This differentiation significantly increases loyalty to Siemens by its most valuable channel partners and has historically produced but-performing sales growth in the elite and select tiers, compared to base-level partners.

As an example of the impact of launching the Market Builder Channel Partner Program, the volume of web-order and EDI (electronic data interface) order transactions by channel partner grew by 230 percent and 180 percent respectively in the first two years of the program. In addition, several of SE&A's channel partners have eliminated competitive product-line sales representation in order to qualify for elite status in the program. Financial incentives, bonuses, favorable business terms, and recognition here all work together to create dif-

S
Energy & Automation

Benefits to Channel Partners

	Residential / Commercial			Industrial / Automation			Motors	
	ELITE	Select	Base	ELITE	Select	Base	ELITE	Select
Sales & Growth Incentive	♦	♦	♦	♦	♦	♦	♦	♦
Promotional Funding	♦	♦	♦	♦	♦	♦	♦	♦
E-Commerce Utilization Bonus	♦	♦	♦	♦	♦	♦		
Startup Services Bonus	♦	♦		♦	♦			
Extended Payment Terms	♦			♦			♦	
Inventory Adjustment Prog.	♦	♦		♦	♦			
Priority Customer Service	♦			♦				
Priority Back-Order Allocation	♦			♦				
Logistics Best-Practice Sharing	♦			♦				
ELITE Eagle Award	♦			♦			♦	
Distributor of Excellence Award	♦			♦			♦	
Special Freight Allowance	♦			♦				

February 2006 -9-

Fig. 9: Benefits to Channel Partners

ferentiation and drive Siemens channel partners to attain higher performance, higher value for Siemens, and higher benefits for themselves (see fig. 9).

E-Business

What is E-Business?

E-business is the electronic transaction of business. The prevailing view is that e-business expands the customer's visibility to the vendor's supply chain processes. A truly effective e-business solution, however, transcends both customer relationship management processes and supply chain processes, creating a total solution for the customer while expanding the market for the vendor. The most familiar example of an electronic medium is an Internet site or store. But electronic data exchange can also occur through EDI (a direct electronic data interface) or B2B solutions (an indirect electronic transfer through an intermediary's standard file structure). Electronic business transactions include, but are not limited to, financial transactions; dissemination of marketing and product collateral; master data synchronization; messaging, tracking, and tracing; electronic mail; video transmissions; pricing; training; and reporting. A subset of e-business is *e-commerce*, which is typically limited to financial transactions such as procurement, order entry, credit requests, pricing requests, returns, and dispute management.

The benefits of e-business are significant for all participants. From the customer's perspective, it allows constant access to vendors and their products and services covering all phases of the supply chain for all functions in the customer's organization. Buyers can research products and compare competitors with a few clicks of the mouse. Purchasing can interface from procurement systems into vendor order-entry systems directly, eliminating duplicate data entry. Logistics can track manufacturing and delivery processes. Accounts payable can electronically receive, load, and store confirmations and invoices instead of using paper. Accounts receivables can submit payment electronically. Most of these capabilities result in reduced costs in the customer's operation, and all of them provide increased flexibility and convenience.

From the vendor's perspective, e-business greatly expands the capability to serve the customer. Marketing concepts and product and service information are accessible to an almost unlimited clientele. Contact information can be collected and leads seamlessly qualified, which expands both the capabilities and capacity of the sales organization. Costs are reduced by allowing the customer more self-service and by eliminating the vendor's duplicate efforts in order management. Orders from customers are accepted electronically without reentry. Order errors are significantly reduced as master data (catalog numbers and pricing) is updated in the customer's procurement system before purchase orders or electronic transfer files are processed. Calls to customer service regarding order status can be eliminated by making order tracking and delivery information directly accessible to the customer. Customer disputes can also be reduced by eliminating one of their most significant causes: pricing errors. Product and service training can be delivered electronically, which expands the training exposure while eliminating costs for instructor travel.

Building an E-Business Solution

E-business projects go through several stages (see fig. 10).

The first step of any e-business project is developing a roadmap. This is particularly important since e-business solutions evolve quickly; technology, processes, competitors, and customers simply change too fast. The roadmap should include beginning-to-end functionality of customer touchpoints with the vendor's organization. Processes for these existing functional areas should be mapped in detail and guided towards a *best-in-class* solution.

A key element at this stage is the determination of the technical infrastructure–both hardware and software. As system availability and performance are critical to good e-business solutions, the technical solution should be responsive, reliable, and redundant, as well as scalable. Organizations should always plan for future expansion, both of the solution's functionality and of the numbers of users.

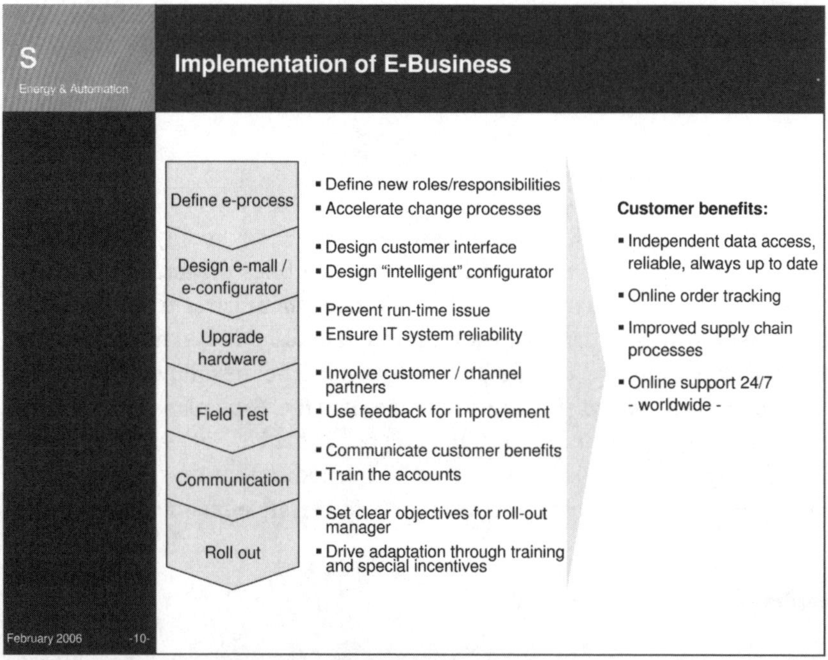

Fig. 10: Implementation of E-Business

A critical requirement for successful deployment is testing. During this time, the functionality provided is tested against the original business requirements. The technical design is tested (integration with other systems) and the entire system is stress-tested for performance. During this testing period, it is crucial to require users–both internal and external–to participate and to agree that expectations are being met. The testing period should allow for any developmental changes in response to the testing feedback.

The final implementation step is the actual deployment to the internal user community and the customer base. A change management plan is designed to support acceptance of the project by the user community. Three significant components of the plan are communication, training, and support. Communication should explain the what, who, why, when, and how of the project and target internal and external users. A training plan should be developed that takes into consideration the complexity of the solution, the roles of the users, and the delivery of the curriculum. E-business solutions can usually be deployed with the aid of some form of web-based training and by providing contextual help pop-ups in the application. However, complex solutions such as product configurators may require more instructor-led support through either web meetings or actual face-to face instruction. In this case, the actual training material should be tested along with the application.

The end of the deployment involves actually turning on the Internet store for all users. From a support standpoint, organizations should plan for the worst;

the application and users should be monitored constantly to ensure system performance and gauge user satisfaction. Resources should be ready to react as necessary until the solution is stable.

Critical Success Factors

As already stated, a project roadmap or plan is the key to any successful e-business implementation. It determines scope, identifies resources, structures timelines, and sets expectations. Another element for success is involving the customer, who will help the organization understand both the market requirements and the available competitive solutions. The usability of the tools selected is *mission critical*. Finally, a change-management plan that includes strong communication, training, and support initiatives is essential to the deployment. One key lesson that has been learned is that on-site support to major customers is the most effective way to accelerate e-business growth.

Results

Fig. 11 shows the strong growth in sales ordered electronically through three methods of input (EDI, B2B, and web in percent of total volume qualified for e-business through fiscal years 2004 and 2005). EDI is a data transfer method that has existed for some time; before the 1990s it was the only electronic order entry method. As a solution, EDI typically addresses high-volume ordering for larger organizations, particularly retail organizations where inventory turns are high. EDI requires a relatively large up-front investment to deploy the technology; doing so leads to relatively low transaction costs thereafter.

In the late 1990s and early 2000s, the Internet store emerged and quickly had a strong impact on the economy of the developed world. This method provides electronic order entry access to virtually anyone connected to the web. Internet stores are generally used by individual consumers or small organizations that do not have enough volume to justify an EDI solution, do not see the benefit of investing in such a solution, or are not technically advanced enough from an IT perspective to support an EDI solution.

The latest solution to hit the market is B2B, which is very similar to EDI but uses a third-party provider as an intermediate platform. Instead of interfacing directly with the vendor as with EDI, a B2B customer interfaces with that platform, which then interfaces with the vendor. With this approach, a customer can use a B2B solution for multiple vendors, the B2B provider directing the vendor-specific part of the data exchange to the appropriate vendor. Investment in a B2B solution is not nearly as high as is the case for EDI, and B2B can handle high volumes of orders. For that reason, B2B is increasingly becoming a solution for smaller organizations that have fairly high volume and that draw on multiple vendors.

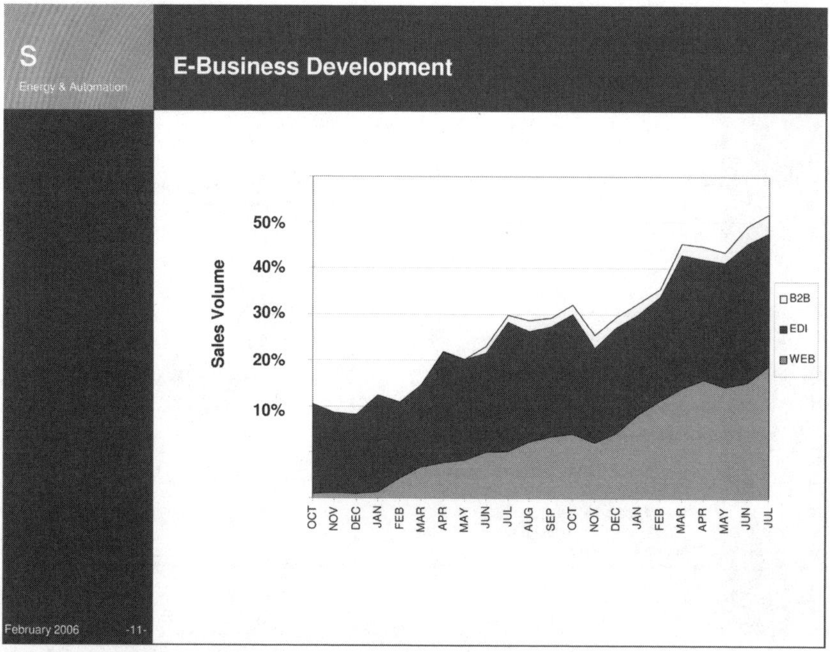

Fig. 11: E-Business Development

From fig. 11, we can see that EDI growth has been relatively low, whereas Internet store or web order entry has dramatically increased. The general view, however, is that B2B will become the solution of growth in the next few years. It is also interesting to note that for SE&A, the volume ordered electronically as percentage of volume qualified has grown from around 10 percent (all EDI) to 50 percent (a mix of 47 percent EDI, 48 percent web, and 5 percent B2B) in a relatively short period. The benefits are obvious: lower cost for order processing, improved quality in the supply chain, and higher customer satisfaction.

Sales Support Mechanisms

Siemens Energy & Automation is a complex matrix organization that operates in a mature but competitive market. It is thus important to the firm to have a very structured and systematic approach to sales planning. Common understanding and good cooperation across the organization is needed to achieve its growth plans; this cooperation is based in turn on various bottom-up and top-down processes.

The regional planning process (RPP) (see fig. 12) is a good example of such a structured approach. When fully implemented, understood, and supported with effective tools, it is a comprehensive platform for growth because it engages all the required units of the organization (horizontal and vertical sales, business units, etc.).

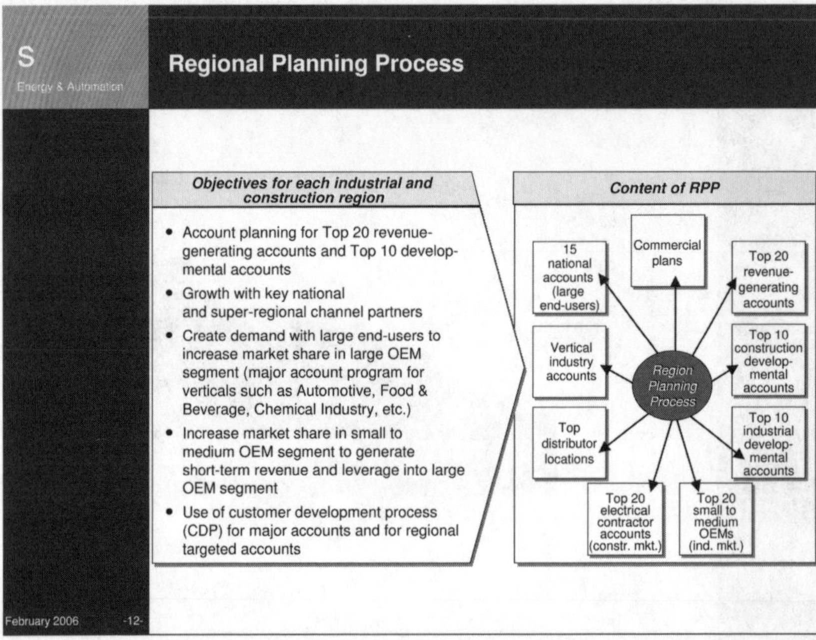

Fig. 12: Regional Planning Process

Essentially, each sales engineer (SE) is tasked with completing comprehensive account plans in a CRM tool (Siebel) for their top established and developmental accounts. Each region is then asked to prepare and present the plan for top accounts in a quarterly review meeting with all key stakeholders. During the meeting, the accounts are reviewed and discussed with the product business units to identify and agree on support requirements (that is, special resource initiatives and investment), in order to secure the identified potential in accordance with the plan.

Each account plan is developed by the account team to ensure that all key customer information on their markets, strategies, objectives, key competitors, and growth initiatives is considered. The planning detail itself focuses on objectives and action plans to increase market share at the account. SE&A develops strategic objectives and determines specific programs and actions against the backdrop of identified potential (fig. 13). The opportunities identified and discussed through the planning process produce an *opportunity pipeline*. Each opportunity is systematically developed by the account manager, through the sales stages (degrees of implementation) to its eventual conclusion (won, lost, abandoned). It is critical at this stage that the CRM tool is effectively used by all parties (sales, marketing, business units, etc.) to ensure coordinated customer response and maximum effectiveness. The CRM tool is the one component that enforces consistency throughout the process; in fact, it is the glue that holds the process together.

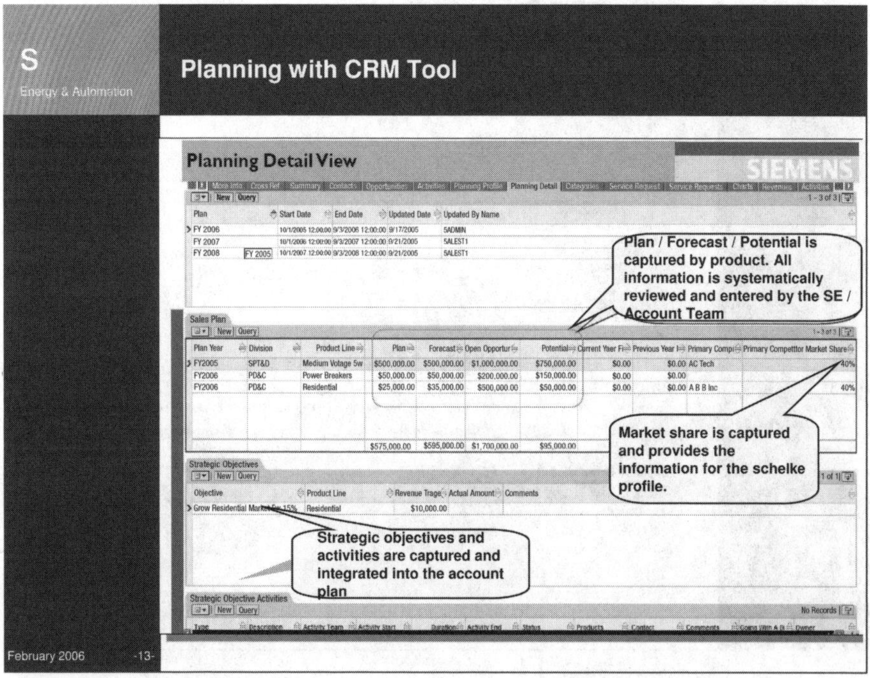

Fig. 13: Planning with CRM Tool

No process is complete unless its impact is measured. Simple and easy-to-use reports are thus essential if one is to effectively manage any process. SE&A has consequently developed dashboard reports to monitor sales behavior (see fig. 14). The focus here is on opportunity values and win-loss ratios, along with sales activities and associated products, thus offering a more complete and balanced view of sales behavior. The reports can be easily accessed, and it is possible to review by division, by region, and by sales manager – even by individual sales engineers. This delivers the transparency needed to manage the process effectively and to maintain flexibility and speed of response. The reports are forwarded to all stake holders across the organization on a weekly basis and are regularly reviewed.

Performance Management and Personnel-Related Processes

KPI and Target Setting

We have stressed that measuring and monitoring all primary sales processes is a basis for assessing progress and influencing individual performance. Monitoring and measuring the growth performance of a sales organization is also critical, with developing growth initiatives through cooperation between sales

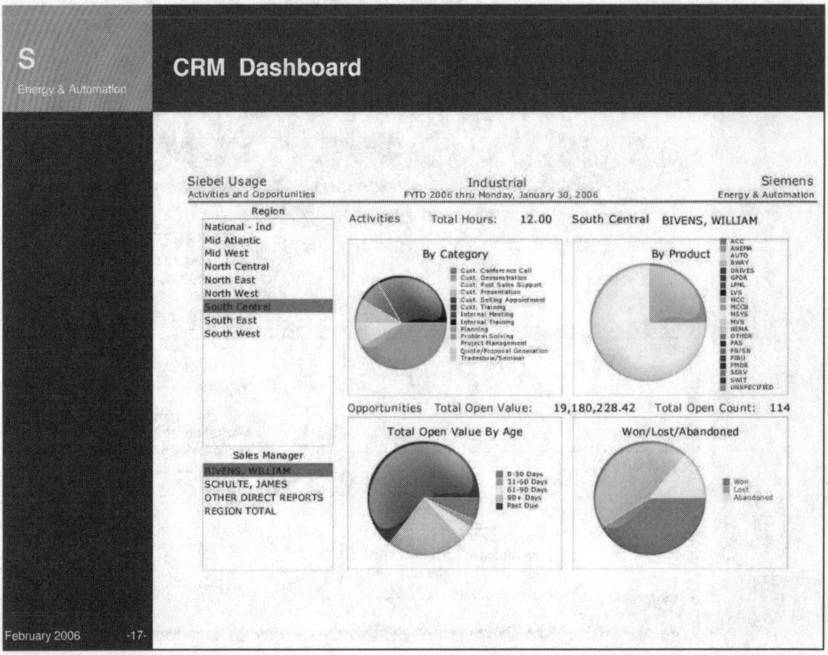

Fig. 14: CRM Dashboard

and product business units constituting a key first step. These initiatives range from product development and introduction to applying additional selling resources in markets where investment is high. Each initiative is monitored and tracked against a goal. A degree of implementation (DI) is assigned to assess whether the goal is on track or at risk of not being achieved.

The summary of these goals (corrected for possible overlap or leakage) makes up the overall stretch target (see fig. 15). Implementation and success of the individual growth initiatives of all the business units is tracked and monitored on a monthly basis.

Performance Monitoring and Review

A core element in successfully addressing the challenges of a complex and global business is the creation and sustaining of a high-performance culture. To that end, SE&A fosters an environment in which expectations are clarified, results and performance acknowledged, and employees encouraged to accept individual responsibility and commit themselves to achieving results.

Employee performance is measured against clearly defined performance objectives that are broken down from top-level management and tied to the organization's balanced scorecard. Employees are also assessed against the Siemens competency framework (see fig. 16). The capabilities noted in this slide

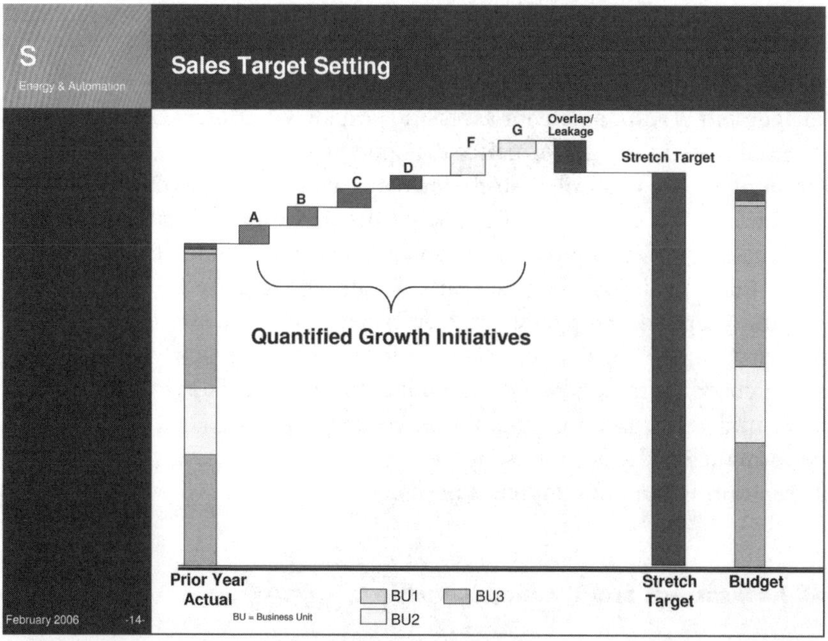

Fig. 15: Sales Target Setting

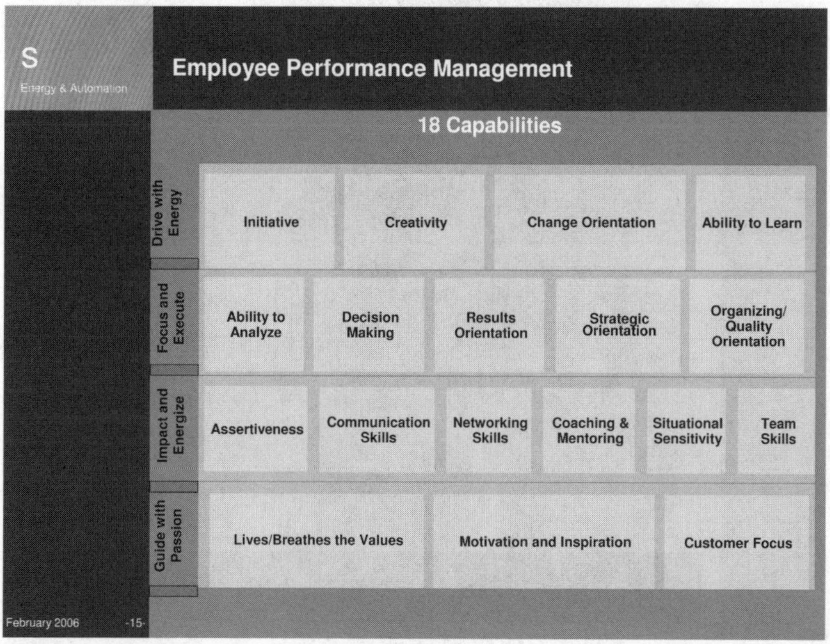

Fig. 16: Employee Performance Management

are consistent across all positions and have been identified as those required for success within the organization. Employee development focuses on these capabilities. An individual development plan is established for every employee.

Management is responsible for assessing employee performance and engaging its employees in the performance dialogue process. Employees are equally held accountable for meeting their performance objectives and communicating their results to their managers. The goal of the performance dialogue process is for managers and employees to establish a common understanding of job responsibilities, to discuss the past year's results, and to agree on objectives for the coming year. The employee's career expectations and the next steps for progress at Siemens are also reviewed. Performance is monitored on a regular basis and coaching is conducted via both informal and formal sessions (for example, mid-year coaching, end-of-year review); another key component of the performance-dialogue process is the identification and communication of employee training and development needs.

Skill Assessment and Training/Coaching

In October 2004, the SE&A sales division launched a learning and development initiative – the "Investing in You" training program (see fig. 17) – based on feedback from the employee survey. The goal of the initiative is to outline the skills and capabilities (knowledge of product and solutions, selling skills,

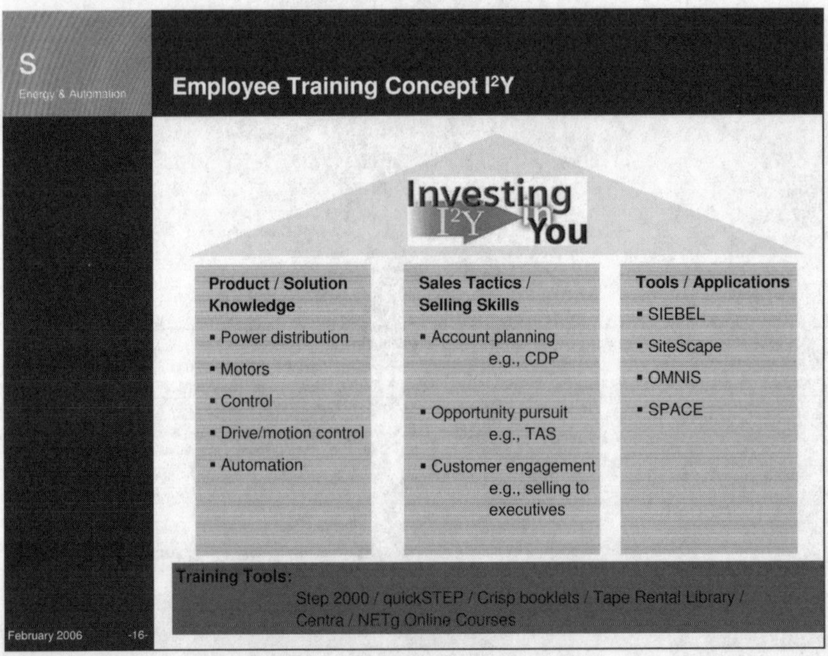

Fig. 17: Employee Training Concept I²Y

tools and applications) required for success in key sales functions. To date, learning maps and career paths have been generated and linked to position descriptions so that employees can more easily research and identify possible career opportunities. The learning map and career path documentation also serves as an aid to managers, who can refer to the information in preparing career discussions with employees.

SE&A is also committed to developing its leadership team. Management courses and leadership programs (for example, the Emerging Leader Program) help develop management talent and keep the talent pipeline full. Leadership development is based on high performance principles and on Siemens leadership-framework capabilities.

Rewards and Recognitions

SE&A's rewards and recognition systems are a key to driving performance and to sustaining a high-performance culture. Incentive compensation plans are used to encourage performance that supports the company's overall business goals. In addition, the company runs an employee recognition program tied to its guiding principles: quality, sales stimulation, asset management, innovation, and cost effectiveness. Merit increases are also tied directly to performance.

Conclusion

Business success results from effective execution of a sales management model, based on systematic processes. To this end, definite roles and responsibilities have to be established. Setting clear goals provides direction and motivation. Leveraging all resources, both internal and external, maximizes the likelihood of success.

Success in a complex international business depends on cross-functional cooperation. This includes all selling functions, channel partners, marketing, and operations. Coordinating global capabilities can also play a key role in gaining customer confidence. A crucial aspect of this success is systematic sales planning based on a high level of customer transparency and a good understanding of the customer's purchasing patterns. It is also very important to develop a clear focus based on segmentation by product, solutions, customers, and go-to-market channels. E-business should be implemented step by step and can only be successful with the right support structure. Many plans fail due to flaws in processes, tools, and communication. At the same time, it cannot be stressed enough that continuous talent development is needed to maintain a competitive edge. And strong leadership is equally important in creating the right environment to strongly motivate the entire sales team.

Interview with Aubert Martin[189]

Olaf Plötner

Mr. Martin, an outsider becoming acquainted with SE&A gains an impression that sales and marketing are structured to a strikingly high degree. Is that in fact typically German, typically American – or typically Siemens?

I believe that rather than reflecting typical qualities of a country or a company, this is determined by the market and the portfolio. The more complex a business is, the more structured its approach has to be. And with over 80,000 products and the very many services we offer to all types of industries and the construction market, our business is very complex. Without clear structures, the overall picture disappears very quickly, and when that happens, companies lose their focus.

Didn't structures already exist before you arrived four years ago?

Structures did indeed exist. At the same time we've done some reorganizing and, above all, more clearly defined responsibilities and communication channels. Beyond that we considered it important to develop the mind-set in the various departments. Earlier, building good personal relations with customers stood at the center of the company's interests. It was important to us that our employees also increasingly worked at gaining a deep understanding of our customers' business. That includes profound know-how regarding both its processes and competitive environment on the one hand, and the specific requirements and challenges of a market segment on the other hand. In the development of the skill-set connected to this, we have offered strong support to employees through additional, career-linked training. Beyond this, we've tied the new demands involved here to our Performance Management System.

Were all employees up to the challenge?

No. Precisely in the initial phase there was a significant change of personnel. But meanwhile things have stabilized. In the sales division, for example, we had to change up to 50 percent of the employees in the beginning. At present, this rate is down to around 10 percent. In addition, extensive training and change in management was necessary.

Were the comprehensive systems for business planning and performance measurement at SE&A developed together with the employees?

Yes; when something like this is only carried out top-down, chances for success are slim. The necessary acceptance by the many good people available

Bringing Technology to Market. Olaf Plötner and Robert E. Spekman

ISBN 978-3-527-50270-7

would simply not be there. To this extent a common effort on the part of all management levels has been needed. This has also been the case for the development of the Balanced Score Card, particularly since we have broken down this approach to the level of the individual business unit. In total, shared development of this instrument lasted a year for us. But the results were rewarding: we're still using the system we developed then. I should note that in order to make the tool's significance clear, after its introduction we arranged monthly business meetings to review the progress of SE&A in total and its individual divisions. This clearly strengthened internal acceptance of the scorecard and the individual accountability.

Weren't there conflicts over goals resulting from the fact that with key account managers and regional sales managers, the use of retail and channel partners, and an e-business platform you are actually using all conceivable sales channels in a parallel manner?

There are indeed conflicts, but many can be managed very well through the definition of clear structures. For example, many supposed conflicts of interest can be avoided from the start through a suitable incentive system. In cases where these conflicts nonetheless exist, clear rules have to be established. These rules have to be openly communicated to everyone and it also doesn't harm things to repeat them now and then. But above all it's important that the rules are followed consistently so that everybody can rely on them. When all those involved know where they stand, then they can much more easily accept things working to their disadvantage. This applies to both employees and sales partners.

But isn't establishing such rules necessarily based on SE&A first clearly defining its own goals?

Yes, you need to know precisely what you want. I am firmly convinced that in sales, focusing is exceptionally important. In the framework of a goal-finding process, this above all includes defining what you *don't* want. Many people in sales find this difficult. We too often find a tendency there to chase after anything promising a sale. But if you want to use any business opportunities offering themselves in one way or another, then your energies are quickly dissipated. Precisely in the case of complicated multi-channel systems like ours, that can lead to confusion and unnecessary complications.

Comprehensive customer understanding, focused goals, clear systems of planning and measuring: are these the central factors in the convincing success of both SE&A and Siemens A&D?

No, the list is missing the most important factor – the discipline to consistently realize what you've planned. I think this often represents the greatest deficit in corporate practice. Often guidelines for strategic planning, operative responsibility and report structures only exist on paper. In everyday operations, you can find a thousand reasons not to follow them. And when that doesn't have any consequences, people stop taking what's been stipulated seriously. The exceptions turn into the rule and people shouldn't be surprised if the company doesn't develop as was expected. For me that's the most important lesson we've learned from the past years, especially regarding sales: what counts is consistent execution. Top management needs to pay special attention to this, and we have to always repeatedly work on it ourselves. But this of course does not mean that developing strategies and structures is unimportant for mastering the future.

What do you consider the most important future challenge for sales and marketing?

While developments have already been very fast paced, the tempo of changes will accelerate even further. Innovative technologies are offering more and more options and advancing globalization is intensifying market dynamism. For this reason understanding customer processes and finding the most attractive value proposition is becoming more and more of a challenge. But we also see a great opportunity for our company here, since in general we would prefer profiling ourselves through the special value of our products and services rather than through low prices. To be sure, in order to use this opportunity we will need a further rise in employees' qualifications. For this reason I see it as a central challenge for our company to remain attractive to the best talents in the future, and to steadily develop their skills in accordance with the needs of the business.

Outlook

Robert E. Spekman

On the surface it would appear that this book is about technology and management's ability to successfully develop, commercialize, and manage innovation through its life cycle. However, that inference would be partially incorrect. The book's title implies that technology is brought to market proactively, and in the context of a B2B marketplace it is often the sales force that does the "heavy lifting," providing the shoulders on which the new technology can be carried. While it would seem that some technology sells itself, this image belies the fact that resistance exists, buyers and influencers are often hard to reach, and the seller often cannot align its own functions and disciplines sufficiently to present a united front to the marketplace. The sad fact is that most innovation fails to achieve any semblance of market penetration adequate to sustaining the product's launch. At the same time, this collective work does offer many examples of success and a number of tools and approaches to raise the probability of market acceptance, along with caution about paths that will self-destruct.

Failure can be attributed to many factors. Work by Utterback,[190] Christensen,[191] and others has suggested a number of explanations of why technology is not easily brought to market. The explanations seem to converge around several themes. One of these is related to the actual attribute set inherent in the technology; it suggests that a relative advantage must exist, that the technology is fairly easy to explain, and that it can be demonstrated so that potential users can get a sense for how it works, and so forth. Other explanations are related to the selling firm's appetite for investing in the technology and doing what is needed to see the innovation through commercialization. Unfortunately, it would seem that the incumbent firm often has too much invested in the current paradigm and finds it difficult to shift adequate resources to the new technology. Ironically, since the newer, often smaller firm has no such baggage, it finds itself willing to take on the task of market development. But it often finds itself short of cash and cannot afford the dedicated sales force to carry the technology to market. Another explanation for the firm's reluctance to innovate is that it can be too close to its existing customers; and since these customers have little incentive to change, innovation is slow.

Still other explanations are related to the structure of the organization and its capacity to effectively execute innovations. Firms that are siloed, have little ability to transfer knowledge across departments, do not see information as a public good, are not likely to innovate easily; nor will they be able to successfully engage other firms to adopt their technology. This problem is exacerbated

Bringing Technology to Market. Olaf Plötner and Robert E. Spekman

ISBN 978-3-527-50270-7

when it unfolds on a global level, since selling in different cultures tends to make issues more complex. In addition, the seller has a number of options for proactively changing the customer's viewpoint, thus making the technology more appealing. For example, the seller can change the set of value-added services that accompany the core product; it is clear that the bundle of such services will change over the product's life cycle. Alternatively, the seller can work with the buyer to reconfigure the value chain, which might alter the existing business model or change the emphasis of the new technology to a different solution or problem.

All said, it would seem that success in bringing technology to market hinges on the quality of execution and doing one's homework. We cannot place enough emphasis on the importance of involving customers early in the process – on truly wiring their voices into all phases of new product development. The effort involved here underscores the importance of clear, defined benefits for solving key customer problems. It is nevertheless fair to ask: who is responsible for these activities? Surely, the importance of marketing research and marketing (product or brand) management is embedded in the process, since data collected from customers feeds directly into product definition and positioning, as well as into segmentation decisions. However, the reality is that many B2B companies spend so little on marketing research that the budget is often equal to rounding errors on an income statement. The fact is that the sales force is the eyes and ears of the firm. The salesperson is the advocate for the customer and the image of the selling firm in the customer's mind. The picture that here emerges is of a sales force playing an increasingly critical role: changes in the sales process, ever more complex demands on the part of the customer, and the dynamics of global competition have all added to the difficulties associated with managing the sales force.

Future Challenges

It is import to recall that innovation is the lifeblood of the firm and that the ability to bring technology to market is what separates winners from losers. With firms seeking a sustainable competitive advantage, we envision a number of challenges that will face the sales leadership. Rather than repeating topics previously mentioned, our wish here is to emphasize those challenges that appear formidable but are manageable: to thus suggest approaches presenting opportunities to those managers willing to change the rules, cognizant of both the risks and the fact that they are far outweighed by the potential rewards.

- Getting close to the customer is not a sales problem; nor is it a marketing problem. It is a responsibility shared by the entire organization. To understand present customer needs and anticipate future needs requires a level of customer intimacy that transcends a single function or department. Firms that can rid their structures of their silos and unite their fiefdoms to serve a common purpose will win the day. Because of sales management's unique position in the buyer-seller relationship, it should be granted a leadership role in the process of getting close to customers.

- Sales often does not have an assigned seat at the strategy table. Rather, with few exceptions it is relegated to a secondary role. For many at the C level, it is viewed merely as a tactical player, not as a high-level strategic contributor. What is apparent is that generally the sales leadership must make a greater investment in improving the quality of its salespeople. Beyond that, it must offer ideas and act as a strategic resource. Skills such as integration, facilitation, and coordination become essential competencies for sales leaders, since they must bring together functions and people both within the firm and between the firm and the customer.
- What is the role of sales? A recent *Business Week* article has suggested that salespeople should be better listeners and problem solvers. Ought we carry this to its extreme and conclude that sales should not sell but advise, consult, and creatively address business problems for consumers? If sales management is not taken seriously within the firm, how will it be taken seriously externally? Recognition of the changing role of salespeople and their management is critical to sales being viewed as a strategic resource.
- Measures of success are often misleading and can lead to mistaken performance. The question arises of the correct performance metrics for the salesperson to emerge as a trusted customer advisor rather than merely as an order taker. Longer selling cycles, team selling, and an inability to equate effort with outcomes has shifted the nature of performance assessment. While still important, actual generated revenue is now being complemented with longer-term objectives such as share of wallet and customer lifetime value. In addition, behavioral indices are increasingly being recognized as important for understanding the process by which the sales force interacts with its customer base. The implications for hiring, motivating, coaching, and promoting are quite profound, since a number of firms have found that making the transition from transactional selling to consultative and enterprise selling is beyond their capacities.
- The use of sales automation tools, customer relationship management software, and enterprise-level solutions has represented a mixed blessing. The sad fact is that these tools often over-promise and under-deliver. However, on a level with graver implications, the tools are expensive to install and maintain, and they run havoc with the firms' operations during early implementation. In this respect, the case treating the approach taken by Siemens A&D to sales automation represents a notable example of success. While much can be learned from this and similar examples, we nevertheless believe in the strong advisability of proceeding with caution, approaching the tools for what they are: no substitute for managerial prerogative but decision aids, pure and simple. We have observed cases where the tools have

taken on a life of their own, metrics having here been developed to encourage the use of a particular sales automation software. This is not a positive development.

- In light of the clear need for a different type of salesperson and new approaches to sales leadership, coaching and mentoring have become key activities for sales management; we believe that many sales leaders are ill prepared for this task. It is important to here keep in mind that sales leaders have often been promoted to their positions because they were strong salespeople – meaning they are likely to have had little experience in management. The essential questions that thus emerge are the following: how are these critical skills in fact learned, and is there a need for a different career path for sales management?
- Given the fact that the "world is becoming flat,"[192] it is likely that it will take the talents of different firms working together as one virtual organization to bring complex solutions to end-use customers – in other words, to bring technology to market. Should this be the case, then the role and that of both the salesperson and management changes once again. Complementing possession of a wide range of skills, the ability to be a systems integrator now emerges as a major factor. In this context, the ideal salesperson is someone who can work across functions, companies, and national borders to weave complex solutions for the customer.

The last point on the above list will hopefully elicit reactions. We cannot help but wonder if selling and the sales function as still understood today has actually run its course. In the market-focused, customer-driven firm, is sales in fact important? If the selling firm is dedicated to solving business problems for customers and playing the role of trusted advisor, selling in the traditional sense seems to fall by the wayside. If this is so, then the traditional business model for hiring, training, motivating, and compensating the salesperson is itself no longer valid. The profile of the salesperson changes, the linkage between sales effort and outcome becoming tenuous at best. Given the new use of teams and partnerships to address customer needs, longer selling cycles, the complexity of the buying process, and the impact of global forces all confirm that the old rules no longer apply. With what then do we replace the sales function, and who will be responsible for filling the present void? We here refer the reader to the July-August 2006 edition of the *Harvard Business Review*, which is dedicated to sales and presents a number of timely papers on its role and significance in a corporate framework. Here it is argued that the current orientation might indeed need to change – that more innovative processes and structures are likely to arise, as is the case in a Schumpeterian framework of "creative destruction."

These challenges are mostly related to internal aspects of sales management and are quite consistent with themes echoed in this book. Nevertheless, challenges facing sales management can also be derived from external factors facing the firm. Interestingly, these external considerations also will have a profound effect on sales management. Among the external challenges are the following:

- The growing globalization of customers and suppliers. When we speak of being world class, it is meant literally. Companies now compete for business on a global basis and shop the world for suppliers that can meet their demands on the same basis. Such issues make considering low-cost country sourcing an important means for achieving a sustainable competitive advantage. Given the rise in global account management, sellers are now discovering that their customers are asking for one invoice for all worldwide purchases issued on the first of the month and that the seller determine the currency of payment. The added administrative complexity required just to service these global accounts is easy to imagine.
- As sellers engage in greater system selling they are forced to better understand the customers' value chain and how value is created in the marketplace. Nevertheless efforts are constantly being made to change the value chain to reap greater margins or else to dis-intermediate members of the chain for the sake of protecting one's position or relationships. The challenge for the seller is to understand enough about its customers to provide solutions reconfiguring the value chain to give them an advantage.
- As sellers try to build closer relationships with buyers, buyers try to narrow the number of firms from which they buy and to form closer, more collaborate ties with select suppliers. While such relationships are a good thing for both parties, the caveat "beware of pirates in partner suits" still holds. Put simply, not all partners are good partners and while the seller might be trusting, the buyer might not be trustworthy. Such a dichotomy can lead to opportunistic behavior, with buyers furthering their interests to the detriment of sellers.

One main purpose of the present volume is to make sense of the uncertainty and tensions that lie ahead. It is clear that the challenges inherent in bringing technology to market are highly complex, spanning problems of management, navigation across functions and organizations, and working in an environment where sales has increasing responsibility but no longer any authority. The situation becomes even more acute in the context of global markets. At the same time, it is important to keep in mind that traditional business transactions may themselves be superseded by the new emerging economic order: an order within which companies collaborate closely to solve problems for customers and customers' customers. We hope that this volume has offered some encouragement in approaching the ongoing transition, and has suggested directions for the emergence of new technologies operating within a new sales model. Although there are many challenges facing the sales community, the challenges are surmountable and the rewards are great.

Acknowledgements

For making the idea of this book a reality, we would like to thank, first and foremost, Friedrich Fröschl, Peter Kuisle, Aubert Martin, Lothar Pauly, Richard Steiner, and Dale Wilson, executives who on various levels were extremely generous with their very limited time.

We would like to also extend a special note of thanks to Joel Golb for his meticulous line editing of each of the chapters, without which their themes and topics would not have crystallized into a unified book. We thank Joel Golb as well for his translation from the German of chapters 1 -4, and the interviews of cases 1-5.

Bringing Technology to Market. Olaf Plötner and Robert E. Spekman

ISBN 978-3-527-50270-7

Notes

Chapter 1

1 *Der Spiegel,* June 7, 2004, 86.
2 Source: WTO, *International Trade Statistics 2001.*
3 Plinke, W., "Grundlagen des Marktprozesses," in Kleinaltenkamp, M., Plinke, W. (eds.), *Technischer Vertrieb: Grundlagen des Business-to-Business Marketing,* 2nd ed., Berlin 2000, 77.
4 Ibid, 86.
5 Narver, J.C., Slater, S.F., "The Effect of a Market Orientation on Business Profitability," *Journal of Marketing* **1990**, 54, 21.
6 Ghemawat, P., *Commitment: The Dynamics of Strategy,* New York 1991, 18.
7 Porter, M.E., *The Competitive Advantage of Nations,* London 1990.
8 Williamson, O.E., *Markets and Hierarchies: Analysis and Antitrust Implications,* New York 1975; idem, *The Economic Institutions* of *Capitalism: Firms, Markets, Relational Contracting,* New York 1985.
9 Söllner, A., "Die neue internationale labor division," in Rese, M., Söllner, A., Utzig, P. (eds.), *Relationship Marketing. Standortbestimmung und Perspektiven,* Berlin 2003, 182.
10 Plinke, W., "Grundlagen des Businessbeziehungsmanagements," in Kleinaltenkamp, M., Plinke, W. (eds.), *Businessbeziehungsmanagement,* Berlin 1997, 1-61.
11 Ibid, 12.
12 Ibid, 44.
13 Deutsche Bundesbank, "Zur Problematik internationaler Vergleiche von Directinvestmentsströmen", *Monatsberichte der Deutschen Bundesbank,* **1997**, 49, 81.
14 Kogut, B., "Joint Ventures: Theoretical and Empirical Perspectives," *Strategic Management Journal,* **1988**, *9*, 319-32; Weder, R., *Joint Venture: Theoretische und empirische Analyse unter besonderer Berücksichtigung der chemischen Industrie der Schweiz,* Grüsch 1989; Yan, A., Gray, B., "Antecedents and Effects of Parent Control," *International Joint Ventures, Journal of Management Studies,* **2001**, *38*, 393-416.
15 Porter, *Competitive Advantage,* 77.
16 Baron, D.P., *Business and its Environment,* 4th ed., Upper Saddle River (NJ) 2003.
17 Daumann, F., *Interessenverbände im politischen Prozeß,* Tübingen 1999, 10.
18 Baron, *Business and its Environment,* 17.
19 Homann, K., "Marktwirtschaft und Unternehmensethik," in Forum für Philosophie Bad Homburg (eds.), *Markt und Moral. Die Diskussion um die Unternehmensethik,* Bern 1994, 109-130.
20 Ibid.
21 Idem, "Der Sinn der Unternehmensethik in der Marktwirtschaft," in Corsten, H. et al (eds.), *Die soziale Dimension der Unternehmung,* Berlin 1991, 112.
22 Carroll, A. B., *Business & Society: Ethics and Stakeholder Management,* 3rd ed., New York 1996, 39.
23 Homann, "Sinn der Unternehmensethik," 108.
24 Idem, "Marktwirtschaft und Unternehmensethik," 118.
25 Source: United Nations, 2005, http://www.un.org/Depts/ptd/global.htm.

Chapter 2

26 Vargo , S.L., Lusch, R.F., "Evolving to New Dominant Logic for Marketing," *Journal of Marketing,* **2004**, *68*, 1–17.
27 Source: *Monitor Study* – "Industrial Services Strategies – The quest for faster growth and higher margins", **2004** – http://www.monitor.com/binary-data/MONITOR_ARTICLES/object/199.PDF, 15.
28 Jacob, F., Plötner, O., Zedler, C., "Competence Commercialization von Industrieunternehmen," Working Paper 17, ESCP-EAP Europäische Wirtschaftshochschule, Berlin 2006.
29 Source: VDMA-Survey, Product-related Services 2001, http://ec.europa.eu/enterprise/standards_policy/services/doc/10–wiechers.pdf.

Bringing Technology to Market. Olaf Plötner and Robert E. Spekman

ISBN 978-3-527-50270-7

30 Darby, M.R., Karni, E., "Free Competition and the Optimal Amount of Fraud," *The Journal of Law and Economics*, **1973**, *17*, 67-88.
31 Plötner, O., *Das Vertrauen des Kunden – Relevanz, Aufbau und Steuerung auf industriellen Märkten*, Wiesbaden 1995.

Chapter 3

32 Kleinaltenkamp, M., "Business-to-Business-Marketing," in *Gabler Wirtschaftslexikon*, Wiesbaden 2000, 602-07.
33 Piller, F. T., "Industrialisierung ist nicht genug," in Frank Piller & Christof Stotko Piller, F. and Stotko, C. (eds.), *Mass Customization und Kundenintegration – Neue Wege zum innovativen Produkt*, Düsseldorf 2003, 29-42.
34 Becker, J., "Massenware fuer die Einzelmännchen," *Sueddeutsche Zeitung*, **2002**, *253*, 6.
35 Cox, M., Alm, R., "The Right Stuff: America's Move to Mass Customization," in National Policy Center Association (eds.), *Policy Report*, **1999**, *225*, 6.
36 Piller, "Industrialisierung"; Reinhard, G., Zäh, M.F. (eds.), *Marktchance Individualisierung*, Berlin/Heidelberg 2003.
37 Piller, "Industrialisierung," 85.
38 Schweikart, J., *Integrative Prozesskostenrechnung: Kundenorientierte Analyse von Leistungen im industriellen Business-to-Business-Bereich*, Wiesbaden 1997.
39 Kleinaltenkamp, M., "Customer-Integration: Kundenintegration als Leitbild für das Business-to-Business-Marketing," in Kleinaltenkamp, M., Fließ, S., Jacob, F. (eds.), *Customer Integration: Von der Kundenorientierung zur Kundenintegration*, Wiesbaden 1996, 13-24.
40 Smith, A., *An Inquiry into the Nature and the Causes of the Wealth of Nations*, London 1876.
41 Donnerberg, R., "Mobiler Datenfunk bringt den Kfz-Service auf täglich 70 000 Auftragstouren," *VDI-Nachrichten*, Jan. 2, 2003.
42 Abele, E., Radtke, Ph., Bitzer, A., "Automobilindustrie im Wandel: Wertschöpfungsarchitekturen der Zukunft," in Reinhardt, Gunther, Zäh, Michael F. (eds.), *Marktchance Individualisierung*, Berlin/Heidelberg 2003, 109-17.
43 Göbel, M., Molfenter, A., "Arbeiten fuer die Riesen," *Die Zeit*, **2002**, *57*; iSuppli Corp. (eds.), iSuppli Forecast 2008.
44 Engelhardt, W. H., Reckenfelderbäumer, M., "Industrielles Service-Management," in Kleinaltenkamp, M., Plinke, W. (eds.), *Markt- und Produktmanagement*, Berlin 1999, 181-280.
45 Wimmer, F., Zerr, K., "Service für Systeme – Service mit System," *Absatzwirtschaft*, **1995**, *39*, 82-87.
46 Monitor Group 2003, http://www.monitor.com/binary-data/MONITOR_ARTICLES/object/199.PDF, exhibit 16, 32-33.
47 Freiling, J., "Pro und Kontra für die Einführung innovativer Betreibermodelle," *Industrie Management*, **2003**, *4*, 32-35; see also Lay, G., *Betreiben statt Verkaufen, Mitteilungen aus der Produktinnovationserhebung*, Karlsruhe 2003, 5.
48 Statistisches Bundesamt Wiesbaden, *Volkswirtschaftliche Gesamtrechnung, 2005*, http://www.destatis.de/basis/d/vgr/vgrtab14.php.
49 Dostal, W., Reinberg, A., "Ungebrochener Trend in der Wissensgesellschaft," *IAB* [Institut für Arbeitsmarkt- und Berufsforschung]-*Kurzbericht*, Nuremberg, *10*, Aug 27, 1999.
50 Verein Deutscher Maschinen- und Anlagenbau (eds.), *Produktbezogene Dienstleistungen im Maschinen und Anlagenbau*, Frankfurt a.M. 2001; IMT Project GmbH (eds.), *Industrial Services: Von der Pflichtübung zum Erlösträger*, Berlin 2002.
51 Kleinaltenkamp, M., "Synergiepotenziale zwischen Industriegüter- und Dienstleistungsforschung," *Die Unternehmung*, **2001**, *55*, 295-306.
52 Kleinaltenkamp, M., Haase, M., "Externe Faktoren in der Theorie der Unternehmung," in Albach, H. et al (eds.), *Die Theorie der Unternehmung in Wissenschaft und Praxis*, Berlin 1999, 167-194.
53 Kleinaltenkamp, M., "Integrativität als Kern einer umfassenden Leistungslehre," in Backhaus, K., Günther, B., Kleinaltenkamp, M., Plinke, W., Raffée, H. (eds.), *Marktleistung und Wettbewerb: Strategische und operative Perspektiven der marktorientierten Leistungsgestaltung. Werner Engelhardt to his 65th Birthday*, Wiesbaden 1997, 83-114.
54 Ibid, 90.
55 Kleinaltenkamp, M., Jacob, F., Leib, R., "Outsourcing Kaufmännischer Dienstleistungen," *Information Management*, **1997**, *12*, 65-72.
56 Ibid, 71.

57 Verband der Automobilindustrie e.V. (eds.), *Annual Report 2002*, Frankfurt a.M. 2002, 106.
58 Duranton, G., Puga, D., "From Sectoral to Functional Urban Specialisation," *CEPR Discussion Paper 2971*, London 2001.

Chapter 4

59 Rappaport A., *Creating Shareholder Value: A Guide for Managers and Investors*, 2nd ed., New York 1998.
60 Atkinson, A., Kaplan, R. S., Young, S. M., *Management Accounting*, 4th ed., Upper Saddle River (NJ) 2004.
61 Bullinger H.-J., "F&E heute: Industrielle Forschung und Entwicklung in der Bundesrepublik Deutschland," *IAO-Studie*, Munich 1990.
62 Hamel, G., Prahalad C.K., "The Core Competence of the Corporation," *Harvard Business Review*, **1990**, 79-91.
63 Oxenfeldt A. R., "The Computation of Costs for Price Decision," *Industrial Marketing Management*, **1977**/*6*, 83-90; Oxenfeldt A. R., "The Differential Method of Pricing," *European Journal of Marketing*, **1979**, *13*, 199-212; Plinke, W., Söllner, A., "Erlösgestaltung im Projektgeschäft," in Plinke, W., Kleinaltenkamp, M. (eds.), Auftrags- und Projektmanagement, Berlin 2006, 117-160 (a translation of this article will be published in 2007).
64 Ellram L. M., Siferd S. P., "Total Cost of Ownership: A Key Concept in Strategic Cost Management Decisions," *Journal of Business Logistics*, **1998**, *19*, 55-84.
65 Alchian, A. A., "Uncertainty, Evolution, and Economic Theory," *Journal of Political Economy*, **1950**, *58*, 211-21; Zintl, R., "Der Homo Oeconomicus: Ausnahmeerscheinung in jeder Situation oder Jedermann in Ausnahmesituationen," *Analyse & Kritik*, **1989**, *11*, 52-69.
66 Rese, M., Herter, V., "Preise und Kosten – Preisbeurteilung im Industriegüterbereich," in Backhaus, K., Voeth, M. (eds.), *Handbuch Industriegütermarketing*, Wiesbaden 2004, 969-88 (a translation of this article will be published in 2006).
67 Ibid.
68 Miller J. G., Vollmann T. E., "The Hidden Factory," *Harvard Business Review*, **1985**, *63*, 142-50.
69 Rese, M., "Entscheidungsunterstützung in Geschäftsbeziehungen mittels Deckungsbeitragsrechnung: Möglichkeiten und Grenzen," in Günter, B., Helm, S. (eds.), *Kundenwert*, 3d ed., Wiesbaden 2006, 276-92.

Chapter 5

70 Sharma, A., Dominguez, L. V., "Channel Evolution: A Framework for Analysis," *Journal of the Academy of Marketing Science*, **1992**, *20*, 1-15; Reid, D., Plank, R.E., "Business Marketing Comes of Age: A Comprehensive Review of the Literature," *Journal of Business-to-Business-Marketing*, **2000**, *7*, 9.
71 Sharma and Dominguez, "Channel Evolution."
72 Napolitano, L., "Customer-Supplier Partnering: A Strategy whose Time has Come," *Journal of Personal Selling and Sales Management*, **1997**, *17*, 1-8.
73 Vargo, S.L., Lusch, R.F., "Evolving to a New Dominant Logic for Marketing," *Journal of Marketing*, **2004**, *68*, 1-23.

Chapter 6

74 Jones, E., Chonko, L., Roberts, J., "Sales for Obsolescence: Perceptions from Sales and Marketing Executives of Individual, Organizational, and Environmental Factors," *Industrial Marketing Management*, **2004**, *33*, 439-56.
75 McMaster, M., "Making Changes Easy for Salespeople," *Sales & Marketing Management*, **2001**, *153*, 74.
76 Zimmer, R., Hugstad, P., "A Contingency Approach to Specializing an Industrial Sales Force," *Journal of Personal Selling and Sales Management*, **1981**, *2*, 27-35.
77 Darmon, R., "Controlling sales force turnover costs through optimal recruiting and training policies," *European Journal of Operational Research*, **2004**, 154, 291-303.
78 Chang, J. et al., "Should You Raid your Rival's Sales Force?" *Sales & Marketing Management*, **2002**, *154*, 8.
79 Patton, W.E., King R. H., "The Use of Human Judgment Models in Sales Force Selection Decisions," *Journal of Personal Selling & Sales Management* **1992**, *12*, 1-14.
80 Johnston, W., Cooper, M., "Industrial Sales Force Selection: Current Knowledge and Needed Research," *Journal of Personal Selling & Sales Management*, **1981**, *2*, 49-57.
81 Pelham, A., "Sales Force Training: What is Needed versus What is Happening," *Proceedings of the Center of Business and Indus-*

trial *Marketing's Thirteenth Annual Academic Workshop*, 2006.
82 Marchetti, M., "What a Sales Call Costs," *Sales and Marketing Management*, 2000, *152*, 80-83.
83 Chang, J. et al: "Happy Sales Force, Happy Return," *Sales & Marketing Management*," 2006, *158*, 2.
84 Darmon, R., "Selecting Appropriate Sales Quota Plan Structures and Quota-Setting Procedures," *Journal of Personal Selling & Sales Management*, 1997, *37*, 1-16.
85 Murphy, W. H., Dacin, P. A., Ford, N. M., "Sales Contest Effectiveness: An Examination of Sales Contest Design Preferences of Field Sales Forces," *Journal of the Academy of Marketing Science*, 2004, *32*, 127-43.
86 Murphy, W. H., "In Pursuit of Short-Term Goals: Anticipating the Unintended Consequences of Using Special Incentives to Motivate the Sales Force," *Journal of Business Research*, 2004, *57*, 1265-75.
87 Boles, J., Donthu, N., Lohtia, R., "Salesperson Evaluation Using Relative Performance Efficiency: The Application of Data Envelopment Analysis," *Journal of Personal Selling & Sales Management*, 1995, *15*, 31-49.
88 Good, D., Stone R., "Selling and Sales Management in Action: Attitudes and Applications of Quotas by Sales Executives and Sales Managers," *Journal of Personal Selling & Sales Management*, 1991, *11*, 57-60.
89 Darmon, R., "Optimal Salesforce Quota Plans under Salesperson Job Equity Constraints," *Canadian Journal of Administrative Sciences*, 2001, *18*, 87-100.
90 Winer, L., "The Effect of Product Sales Quotas on Sales Force Productivity," *Journal of Marketing Research*, 1973, *10*, 180-83.
91 Jobbar, D., Hooley, G., Shipley, D., "Organizational Size and Salesforce Evaluation Practices," *Journal of Personal Selling & Sales Management*, 1993, *13*, 37-49.
92 Boles, Donthu, and Lohtia, "Salesperson Evaluation."

Chapter 7

93 Kotler, P., Armstrong, G., *Principles of Marketing*, 10th ed., Upper Saddle River (NJ) 2001, 624-60.
94 Reid, D. A., Plank, R. E., "Business Marketing Comes of Age: A Comprehensive Review of the Literature," *Journal of Business-to-Business Marketing*, 2000, *7*, 96.
95 Gilliland, D.I., Johnston, W.J., "Toward a Model of B2B Marketing Communications Effects," *Industrial Marketing Management*, 1997, *26*, 15-29.
96 Hutt, M.D., Speh, T.W., "A Strategic View of Industrial and Organizational Markets," in idem, *Business Marketing Management: A. Strategic View of Industrial and Organizational Markets*, 8th ed., Mason (OH) 2004, 59-83.
97 Homburg, Ch., Kuester, S., "Towards an Improved Understanding of Industrial Buying Behavior: Determinants of the Number of Suppliers," *Journal of Business-to-Business Marketing*, 2001, *8*, 5-33.
98 Fließ, S., Kleinaltenkamp, M., "Blueprinting the Service Company: Managing Service Processes Efficiently," *Journal of Business Research*, 2004, *4*, 392-405.
99 Sheth, J. N., Parvatiyar, A., "Evolving Relationship Marketing into a Discipline," *Journal of Relationship Marketing*, 2002, *1*, 3-16.
100 Low, J., Blois, K., "The Evolution of Generic Brands in Industrial Markets," *Industrial Marketing Management*, 2002, *31*, 385-92.,
101 Kotler, P., *Marketing Management: Analysis, Planning, and Control*, Englewood Cliffs (NJ) 1967. A more recent approach can be found in Taghian, M., Shaw, R., "The Marketing Audit and Business Performance: A Review and Research Agenda," *Proceedings of the Australian and New Zealand Marketing Academy Conference, 1998*, 2557-71.
102 Shapiro, B.P., Rangan, V. K., John J. S., "Staple Yourself to an Order," *Harvard Business Review*, 1992, *71*, 113-22.
103 Bruhn, M., Schmidt, S.J., Tropp, J. (eds.), *Integrierte Kommunikation, Theorie und Praxis*, Wiesbaden 2000.
104 Dwyer, R.F., Tanner, J.F., *Business Marketing: Connecting Strategy, Relationships and Learning*, 2d ed., New York 2002.
105 Ibid.
106 Bruhn, M., "Integrierte Kommunikation und Relationship Marketing," Bruhn, M., Schmidt S.J., Tropp J. (eds.), *Integrierte Kommunikation in Theorie und Praxis*, Wiesbaden 2000, 5–20.
107 Möller, K., Rajala, A., "Organizing Marketing in Industrial High-Tech Firms," *Industrial Marketing Management* 1999, *28*, 521–35.
108 Sheth, J.N., Sisodia, R. S., "Marketing Productivity: Issues and Analysis," *Journal of Business Research*, 2002, *55*, 349-62.

109 Ingam, T.N., LaForge, R.W., Leigh, T.W., "Selling in the New Millennium: A Joint Agenda," *Industrial Marketing Management*, 2002, *31*, 559-67.

110 Nonaka, I., et al., "Hypertext Organization for Accelerating Organizational Knowledge Creation," *Diamond Harvard Business Review*, August/September 1992, 12–22; Jacob, F., "Auftragsmanagement," in Kleinaltenkamp, M., Plinke, W., et al (eds.), *Auftrags- und Projektmanagement*, Berlin/ Heidelberg/New York, 1998, 1–67.

111 Ingram et al., ibid.

Chapter 8

112 Stanton, W.J., Spiro, R., *Management of a Sales Force*, 10th ed., New York 1998.

113 Ingram, T. N., et al., "New Directions in Sales Leadership Research," *Journal of Personal Selling & Sales Management*, 2005, *25*, 138–54.

114 Brown, S. P., et al., "Adapting Motivation, Control, and Compensation Research to a New Environment," ibid., 155–67.

115 Chonko, L. B., Jones, E., "The Need for Speed: Agility Selling," ibid., 371.

116 Arnett, D. B., Macy, B. A., Wilcox, J.B., "The Role of Adapting Motivation in Supplier-Buyer Relationships," ibid., 27–42.

117 Brown et al., "Adapting Motivation, Control, and Compensation."

118 Dubinsky, A. J., *Sales Training: An Analysis of Field Sales Techniques*, Ann Arbor 1980.

119 Moncrief, W. C., Marshall, G.W., "The Evolution of the Seven Steps of Selling," *Industrial Marketing Management*, 2005, *34*, 13–22.

120 Zoltners, A.A., Sinha, P., Lorimer, S.E., *Sales Force Design for Strategic Advantage*, New York 2006.

121 Tanner, J. F. Jr., Shipp, S., "Sales Technology within the Salesperson's Relationships: A Research Agenda," *Industrial Marketing Management*, 2005, *34*, 305–12.

122 Gohmann, S. F., Guan,J., Barker, R. M., Faulds, D.J., "Perceptions of Sales Force Automation: Differences between Sales Force and Management," ibid., 337–43.

123 Ko, D.-G., Dennis, A. R., "Sales Force Automation and Sales Performance: Do Experience and Expertise Matter?" *Journal of Personal Selling & Sales Management*, 2004, *24*, 311–22.

124 Kale, S. H., "CRM Failure and the Seven Deadly Sins," *Marketing Management*, 2004, *13*, 42–46.

125 Fujitsu Consulting (formerly DMR Consulting), "Strong Leadership, Strategic Vision and Effective Metrics Key to Successful CRM: DMR Consulting Survey Reveals," press release, 26 Feb. 2002; see http://www.fujitsu.com/us/news/pr/fc–20020226-01.html.

126 Bush, A. J., Moore, J. B., Rocco, R., "Understanding Sales Force Automation Outcomes: A Managerial Perspective," *Industrial Marketing Management*, 2004, *34*, 369–77.

127 Arnett, et al., "The Role of Adapting Motivation."

128 Rouzies, D., et al., "Sales and Marketing Integration: A Proposed Framework," *Journal of Personal Selling & Sales Management*, 2005, *25*, 113–22.

129 Cron, W. L., et al., "Salesperson Selection, Training, and Development: Trends, Implications, and Research Opportunities," ibid., 123–36.

130 Ingram, T. N., "Future Themes in Sales and Sales Management: Complexity, Collaboration, and Accountability," *Journal of Marketing Theory and Practice*, 2004, *12*, 18–28.

131 MacKenzie, S. B., Podsakoff, P. M., Rich, G. A., "Transformational and Transactional Leadership and Salesperson Performance," *Journal of the Academy of Marketing Science*, 2001, *29*, 115–34.

132 Ibid.

133 Ahearne, M., Jelinek, R., Rapp, A., "Moving Beyond the Direct Effect of SFA Adoption on Salesperson Performance: Training and Support as Key Moderating Factors," *Industrial Marketing Management*, 2005, *34*, 379–88.

134 Galea, C., "The 2004 Compensation Survey," *Sales and Marketing Management*, 2004, *156*, 28–34.

135 Amyx, D., Alford, B. L., "The Effects of Salesperson Need For Achievement and Sales Manager Leader Reward Behavior," *Journal of Personal Selling &Sales Management*, 2005, *25*, 345–59.

136 Galea, "The 2004 Compensation Survey."

137 Baldauf, A., Cravens, D. W., Piercy, N. F., "Sales Management Control Research: Synthesis and an Agenda for Future Research," *Journal of Personal Selling & Sales Management*, 2005, *25*, 7–26.

138 Ibid.

139 Anderson, E., Oliver, R., "Perspectives on Behavior-Based versus Outcome-Based Salesforce Control Systems," *Journal of Marketing*, 1987, *51*, 76–88.
140 Jaworski, B. J., "Towards a Theory of Marketing Control: Environmental Context, Control Types, and Consequences," ibid., 1988, *52*, 23–39.
141 Brown et al., "Adapting Motivation, Control, and Compensation."
142 Churchill, G. A. Jr., et al., "The Determinants of Sales Force Performance: A Meta-Analysis," *Journal of Marketing Research*, 1985, *22*, 103–18.
143 Ibid.
144 Harris, E. G., Mowen, J. C., Brown T. J., "Re-examining Salesperson Goal Orientations: Personality Influencers, Customer Orientation, and Work Satisfaction," *Academy of Marketing Science*, 2005, *33*, 19–35.
145 Amyx and Alford, "The Effects of Salesperson Need."
146 Sullivan, T., "Evaluating Sales Training Programs: Determining the Effectiveness of Sales Training Programs," Siebel Systems White Paper, San Mateo (CA) 2000.
147 Sujan, H., Weitz, B.A., Sujan, M., "Increasing Sales Productivity by Getting Salespeople to Work Smarter," *Journal of Personal Selling & Sales Management*, 1988, *8*, 9–19.
148 Brashear, T. G., et al., "An Exploratory Study of the Relative Effectiveness of Different Types of Sales Force Mentors," ibid., 2006, *26*, 7–18.
149 Lankau, M.J., Scandura, T.A., "An Investigation of Personal Learning in Mentoring Relationships," *Academy of Management Journal*, 2002, *45*, 779-790.
150 Slater, S. F., Olson, E.M., "Strategy Type and Performance: The Influence of Sales Force Management," *Strategic Management Journal*, 2000, *21*, 813-29.
151 Zablah, A. R., Johnston, W. J., Bellenger, D.N., "Transforming Partner Relationships through Technological Innovation," *Journal of Business & Industrial Marketing*, 2005, *20*, 355–63.

Chapter 9

152 Anderson and Oliver, "Perspectives," 76-88.
153 Challagalla, G. N., Shervani, T. A., "Dimensions and Types of Supervisory Control: Effects on Salesperson Performance and Satisfaction," *Journal of Marketing*, 1996, *60*, 89-105.
154 Jaworski, B. J., Stathakopoulos, V., Krishnan, H. S., "Control Combinations in Marketing: Conceptual Framework and Empirical Evidence", ibid., 1993, *57*, 57-69.
155 Oliver, R. L., Anderson, E., "An Empirical Test of the Consequences of Behavior and Outcome-Based Sales Control Systems," ibid., 1994, *58*, 53-67.
156 Lusch, R. F., Jaworski, B. J., "Management Controls, Role Stress, and Retail Store Manager Performance," *Journal of Retailing*, 1991, *67*, 397-419.
157 Oliver and Anderson, "An Empirical Test."
158 Jaworski, Stathakopoulos, and Krishnan, "Control Combinations in Marketing."
159 Oliver and Anderson, "An Empirical Test."
160 Cron, W. L., Levy, M., "Sales Management Performance Evaluation: A Residual Income Perspective," *Journal of Personal Selling & Sales Management*, 1987, *7*, 57–66.
161 Behrman, D. N., Perreault, W. D., "Measuring the Performance of Industrial Salespersons," *Journal of Business Research*, 1982, *10*, 355-70.
162 Weitz, B. A., "Effectiveness in Sales Interactions: A Contingency Framework," *Journal of Marketing*, 1981, *41*, 85-103.
163 Chonko, L. B., et al., "The Role of the Environmental Turbulence, Readiness for Change, and Salesperson Learning in the Success of Sales Force Change," *Journal of Personal Selling & Sales Management*, 2002, *22*, 227-45.
164 Cravens, D. W., et al., "Behavior-Based and Outcome-Based Sales Force Control Systems," *Journal of Marketing*, 1993, *57*, 47-59.

Case 1

165 All quotations of T-Systems are based on personal interviews conducted through the European School of Management and Technology (esmt).
166 Bell, M., Caldwell, F., "IT Management Scenario for 2015: IT Investment Drivers Change in Four Possible Futures," Gartner Research, November 18, 2005; see also: abstract at http://www.Gartner.com.
167 2004 Ittoolbox Sourcing Survey, http://sap.ittoolbox.com/documents/research/sourcinsurvey.asp?survey=sourcing–survey&i.

168 Member of the Board of Management, Business Customers, Deutsche Telekom, and Chief Executive Officer, T-Systems.

Case 2

169 This case is factual. However, all customer and contractual information, e.g., the number of machines and scope of the contract, has been changed.

170 The web offset system is based on an offset printing press for which the paper is delivered on a reel and one or several complete paper webs are printed on both sides. The paper web is fed over a reel changer to the printing unit. This component winds the reel and exchanges finished reels. Web offset systems are used for the industrial printing of high volumes of newspapers, magazines, advertising supplements, and other printing materials. Offset printing is the most widespread printing method today. It is a planographic printing method, i.e. the printing and non-printing areas are located on a plane on the printing plate but differ in their chemical and physical characteristics. The ink-friendly points on the printing plate are applied with ink. The ink is then transferred (offset) from the printing plate to a rubber blanket and then onto the paper.

171 In the sheet-fed offset system, a feeder unit feeds the printing material from piles of paper sheets to the printing units; the sheets are printed on one or both sides depending on the system's design. Further processing is then carried out in separate working steps.

172 FAZ Net, Nov. 21, 2005.

173 Interview B. Gögdün and O. Plötner, April 27, 2006.

174 Bearer rings are made of reinforced steel and are mounted on the surface of the plate and rubber-blanket cylinder of an offset printing system. They roll off each other, thus reducing vibration created when the tensioning devices for the printing plates and rubber blankets (tensioning channels) meet. These high-precision components need to be checked regularly for good functioning and exact tensioning in order to ensure that printing quality remains high.

175 A folding unit is a component of a web offset system. It cuts and folds the printed paper sheet to produce the end product. The products are then either ready for direct use, for example, in the form of newspapers, or the semi-finished products are bound as books.

176 Executive Vice President of Sales, MAN Roland Druckmaschinen AG.

Case 3

177 Manager of Strategic Marketing and Controlling at Leistritz Extrusionstechnik GmbH.

Case 4

178 EITO (European Information Technology Observatory) in cooperation with IDC (International Data Corporation).

179 See http://nobelprize.org/nobel–prize/physics/laurates/; Leo Esaki (1973 – $^1/_4$ prize); Gerd Binnig (1986 – $^1/_4$ prize); Heinrich Rohrer (1986 – $^1/_4$ prize); K. Alexander Müller (1987 – $^1/_2$ prize); Johannes Georg Bednorz (1987 – $^1/_2$ prize).

180 *IBM Annual Report 2001.*

181 *IBM Global Services: A Brief History,* IBM Corporate Archives, May 2002, 4.

182 Ibid.

183 *IBM Annual Report 1997; IBM Annual Report 2002.*

184 See www.research.ibm.com/about/work_-customers.shtml.

185 Schlender, B., "How Big Blue Is Turning Geeks into Gold," *Fortune*, 2003, *147*, 11.

186 *IBM Annual Report 2005.*

187 Source: Analysts' Meeting, September 7, 1999, Dr. Friedrich Fröschl.

188 Former president and CEO of Siemens Business Services GmbH (SBS).

Case 5

189 President and CEO of Siemens Energy & Automation, Inc.

Outlook

190 Utterback, J. M., *Mastering the Dynamics.*

191 Christensen, C., *Innovator's Dilemma*, Cambridge (MA), 1997.

192 Friedman, T., *The World is Flat.*

The Authors

Christoph Burger heads the Practice Group Telecommunications, Transport, Utilities at esmt customized solutions. He has research interests in customer management, business scenario planning, and business plan assessment.

Franziska Frank heads Business Development Russia, Eastern Europe, and China at esmt customized solutions.

Bülent Gögdün is Program Director at esmt and is responsible for business development in Turkey.

Frank Jacob holds the Chair of Marketing at the Berlin campus of ESCP-EAP European School of Management. His fields of interest are industrial marketing and services marketing.

Wesley J. Johnston is CBIM RoundTable Professor of Marketing at J. Mack Robinson College of Business, Georgia State University, Atlanta. His research focus is on sales force management and B2B-marketing.

Michael Kleinaltenkamp is Professor of Marketing, particularly B2B-Marketing, at Freie Universität Berlin. His research focuses on service marketing and relationship management.

Martin Kupp is Program Director at esmt customized solutions. His research interest focus on business strategy in technology markets, strategic and technology-enabled innovation, and sales strategy on B2B-markets.

Aubert Martin is President and Chief Executive Officer of Siemens Energy & Automation, Inc. He has executive responsibility for the company's strategic direction, operating performance, and marketplace success. He is a member of the company's Board of Directors and a member of the Board of Voest-Alpine Industries, Inc.

Benjamin Moor has spent five years consulting in a variety of technology markets. He is currently a candidate for a Masters in Business Administration at the Darden School of Business, University of Virginia.

Bringing Technology to Market. Olaf Plötner and Robert E. Spekman

ISBN 978-3-527-50270-7

Wulff Plinke is Founding Dean of esmt. Prior to this position he was Professor of Industrial Marketing Management at Universität Hannover, at Freie Universität Berlin, and at Humboldt-Universität Berlin.

Olaf Plötner is Managing Director of esmt (European School of Management and Technology) customized solutions and a lecturer in Accounting at Freie Universität Berlin. His research focuses on customer relationship management, business management, communication, and marketing strategy.

Mario Rese is Professor for Business Administration and Marketing at Ruhr-Universität Bochum. He has research interests in competitive strategy, customer analysis, and marketing controlling.

Brian N. Rutherford is an Assistant Professor of Business Administration at Richards College of Business, University of West Georgia. His research interests are in the areas of inter and intra organizational selling and sales management.

Arun Sharma is Professor of Marketing at the School of Business Administration, University of Miami, Florida.

Jagdith N. Sheth holds the Charles H. Kellstadt Chair of Marketing in the Goizueta Business School at Emory University, Atlanta. He has published widely in different areas of marketing and business strategy.

Albrecht Söllner is Professor for International Management at the European University Viadrina, Frankfurt (Oder). His research interests include new institutional economics, path dependency, business relationships, and the international division of labor.

Robert E. Spekman is Tayloe Murphy Professor of Business Administration at the Darden School of Business, University of Virginia. He is an internationally recognized authority on B2B-marketing, supply chain / channels management, and strategic alliances.

Dale Wilson is Vice President, Sales Division of Siemens Energy & Automation. He directs activities involving sales of the company's industrial products, power distribution, and residential businesses.

Index

Bringing Technology to Market. Olaf Plötner and Robert E. Spekman

ISBN 978-3-527-50270-7